Charles Fort

NEW LANDS VOL II

Édition : BoD · Books on Demand, 31 avenue Saint-Rémy, 57600 Forbach, bod@bod.fr
Impression : Libri Plureos GmbH, Friedensallee 273, 22763 Hamburg (Allemagne)
ISBN : 978-2-3226-1355-7
Dépôt légal : Avril 2025

NEW LANDS VOL II

Charles Fort

PART II

13

June, 1801—a mirage of an unknown city. It was seen, for more than an hour, at Youghal, Co. Cork, Ireland—a representation of mansions, surrounded by shrubbery and white palings—forests behind. In October, 1796, a mirage of a walled town had been seen distinctly for half an hour at Youghal. Upon March 9, 1797, had been seen a mirage of a walled town.

Feb. 7, 1802—an unknown body that was seen, by Fritsch, of Magdeburg, to cross the sun (*Observatory*, 3-136).

Oct. 10, 1802—an unknown dark body was seen, by Fritsch, rapidly crossing the sun (*Comptes Rendus*, 83-587). Between 10 and 11 o'clock, morning of Oct. 8, 1803, a stone fell from the sky, at the town of Apt, France. About eight hours later, "some persons believed that they felt an earthquake" (*Rept. B. A.*, 1854-53).

Upon August 11, 1805, an explosive sound was heard at East Haddam, Connecticut. There are records of six prior sounds, as if of explosions, that were heard at East Haddam, beginning with the year 1791, but, unrecorded, the sounds had attracted attention for a century, and had been called the "Moodus" sounds, by the Indians. For the best account of the "Moodus" sounds, see the *Amer. Jour. Sci.*, 39-339. Here a writer tries to show the phenomena were subterranean, but says that there was no satisfactory explanation.

Upon the 2nd of April, 1808, over the town of Pignerol, Piedmont, Italy, a loud sound was heard: in many places in Piedmont an earthquake was felt. In the *Rept. B. A.*, 1854-68, it is said that aerial phenomena did occur; that, before the explosion, luminous objects had been seen in the sky over Pignerol, and that in several of the communes in the Alps aerial sounds, as if of innumerable stones colliding, had been heard, and that quakes had been felt. From April 2 to April 8, forty shocks were recorded at Pignerol; sounds like cannonading were heard at Barga. Upon the 18th of April, two detonations were heard at La Tour, and a luminous object was seen in the sky. The supposition, or almost absolute belief of most persons is that from the 2nd to the 18th of April this earth had moved far in its orbit and was rotating so that, if one should explain that probably meteors had exploded here, it could not very well be thought that more meteors were continuing to pick out this one point upon a doubly moving planet. But something was specially related to this one local sky. Upon the 19th of April, a stone fell from the sky at Borgo San Donnino, about 40 miles east of Piedmont (*Rept. B. A.*, 1860). Sounds like cannonading were heard almost every day in this small region. Upon the 13th of May, a red cloud such as marks the place of a meteoric explosion was seen in the sky. Throughout the rest of the year, phenomena that are now listed as "earthquakes" occurred in Piedmont. The last occurrence of which I have record was upon Jan. 22, 1810.

Feb. 9, 1812—two explosive sounds at East Haddam (*Amer. Jour. Sci.*, 39-339).

July 5, 1812—one explosive sound at East Haddam (*Amer. Jour. Sci.*, 39-339).

Oct. 28, 1812—"phantom soldiers" at Havarah Park, near Ripley, England (*Edinburgh Annual Register*, 1812-II-124). When such appearances are explained by meteorologists, they are said to be displays of the aurora borealis. Psychic research explains variously. The physicists say that they are mirages of troops marching somewhere at a distance.

Night of July 31, 1813—flashes of light in the sky of Tottenham, near London (*Year Book of Facts*, 1853-272). The sky was clear. The flashes were attributed to a storm at Hastings, 65 miles away. We note not only that the planet Mars was in opposition at this time (July 30), but in one of the nearest of its oppositions in the 19th century.

Dec. 28, 1813—an explosive sound at East Haddam.

Feb. 2, 1816—a quake at Lisbon. There was something in the sky. Extraordinary sounds were heard, but were attributed to "flocks of birds." But six hours later something was seen in the sky: it is said to have been a meteor (*Rept. B. A.*, 1854-106).

Since the year 1788, many earthquakes, or concussions that were listed as earthquakes, had occurred at the town of Comrie, Perthshire, Scotland. Seventeen instances were recorded in the year 1795. Almost all records of the phenomena of Comrie start with the year 1788, but, in Macara's *Guide to Creifi*, it is said that the disturbances were recorded as far back as the year 1597. They were slight shocks, and until the occurrence upon Aug. 13, 1816, conventional explanations, excluding all thought of relations with anything in the sky, seemed adequate enough. But, in an account in the London *Times*, Aug. 21, 1816, it is said that, at the time of the quake of August 13, a luminous object, or a "small meteor," had been seen at Dunkeld, near Comrie; and, according to David Milne (*Edin. New Phil. Jour.*, 31-110), a resident of Comrie had reported "a large luminous body, bent like a crescent, which stretched itself over the heavens."

There was another quake in Scotland (Inverness) June 30, 1817. It is said that hot water fell from the sky (*Rept. B. A.*, 1854-112).

Jan. 6, 1818—an unknown body that crossed the sun, according to Loft, of Ipswich; observed about three hours and a half (*Quar. Jour. Roy. Inst.*, 5-117).

Five unknown bodies that were seen, upon June 26, 1819, crossing the sun, according to Gruithuisen (*An. Sci. Disc.*, 1860-411). Also, upon this day, Pastorff saw something that he thought was a comet, which was then somewhere near the sun, but which, according to Olbers, could not have been the comet (Webb, *Celestial Objects*, p. 40).

Upon Aug. 28, 1819, there was a violent quake at Irkutsk, Siberia. There had been two shocks upon Aug. 22, 1813 (*Rept. B. A.*, 1854-101). Upon April 6, 1805, or March 25, according to the Russian calendar, two stones had fallen from the sky at Irkutsk (*Rept. B. A.*, 1860-12). One of these stones is now in the South Kensington Museum, London. Another violent shock at Irkutsk, April 7, 1820 (*Rept. B. A.*, 1854-128).

Unknown bodies in the sky, in the year 1820, February 12 and April 27 (*Comptes Rendus*, 83-314).

Things that marched in the sky—see Arago's *Œuvres*, 11-576, or *Annales de Chimie*, 30-417—objects that were seen by many persons, in the streets of Embrun, during the eclipse of Sept. 7, 1820, moving in straight lines, turning and retracting in the same straight lines, all of them separated by uniform spaces.

Early in the year 1821—and a light shone out on the moon—a bright point of light in the lunar crater Aristarchus, which was in the dark at the time. It was seen, upon the 4th and the 7th of February, by Capt. Kater (*An. Reg.*, 1821-689); and upon the 5th by Dr. Olbers (*Mems. R. A. S.*, 1-159). It was a light like a star, and was seen again, May 4th and 6th, by the Rev. M. Ward and by Francis Bailey (*Mems. R. A. S.*, 1-159). At Cape Town, nights of Nov. 28th and 29th, 1821, again a star-like light was seen upon the moon (*Phil. Trans.*, 112-237).

Quar. Jour. Roy. Inst., 20-417:

That, early in the morning of March 20, 1822, detonations were heard at Melida, an island in the Adriatic. All day, at intervals, the sounds were heard. They were like cannonading, and it was supposed that they came from a vessel, or from Turkish artillery, practicing in some frontier village. For thirty days the detonations continued, sometimes thirty or forty, sometimes several hundred, a day.

Upon April 13, 1822, it seems, according to description, that clearly enough was there an explosion in the sky of Comrie, and a concussion of the ground —"two loud reports, one apparently over our heads, and the other, which followed immediately, under our feet" (*Edin. New Phil. Jour.*, 31-119).

July 15, 1822—the fall of perhaps unknown seeds from perhaps an unknown world—a great quantity of little round seeds that fell from the sky at Marienwerder, Germany. They were unknown to the inhabitants, who tried to cook them, but found that boiling seemed to have no effect upon them. Wherever they came from, they were brought down by a storm, and two days later, more of them fell, in a storm, in Silesia. It is said that these corpuscles were identified by some scientists as seeds of*Galium spurium*, but that other scientists disagreed. Later more of them fell at Posen, Mecklenburg. See *Bull. des Sci.* (*math., astro., etc.*) 1-1-298.

Aug. 19, 1822—a tremendous detonation at Melida—others continuing several days.

Oct. 23, 1822—two unknown dark bodies crossing the sun; observed by Pastorff (*An. Sci. Disc.*, 1860-411).

An unknown, shining thing—it was seen, by Webb, May 22, 1823, near the planet Venus (*Nature*, 14-19).

More unknowns, in the year 1823—see *Comptes Rendus*, 49-811 and Webb's *Celestial Objects*, p. 43.

February, 1824—the sounds of Melida.

Upon Feb. II, 1824, a slight shock was felt at Irkutsk, Siberia (*Rept. B. A.,*

1854-124). Upon February 18, or, according to other accounts, upon May 14, a stone that weighed five pounds, fell from the sky at Irkutsk (*Rept. B. A.*, 1860-70). Three severe shocks at Irkutsk, March 8, 1824 (*Rept. B. A.*, 1854-124).

September, 1824—the sounds of Melida.

At five o'clock, morning of Oct. 20, 1824, a light was seen upon the dark part of the moon, by Gruithuisen. It disappeared. Six minutes later it appeared again, disappeared again, and then flashed intermittently, until 5:30 A.M., when sunrise ended the observations (*Sci. Amer. Sup.*, 7-2712). And, upon Jan. 22, 1825, again shone out the star-like light of Aristarchus, reported by the Rev. J. B. Emmett (*Annals of Philosophy*, 28-338).

The last sounds of Melida of which I have record, were heard in March, 1825. If these detonations did come from the sky, there was something that, for at least three years, was situated over, or was in some other way specially related to, this one small part of this earth's surface, subversively to all supposed principles of astronomy and geodesy. It is said that, to find out whether the sounds did come from the sky, or not, the Prêteur of Melida went into underground caverns to listen. It is said that there the sounds could not be heard.

14

AND our own underground investigations—and whether there is something in the sky or not. We are in a hole in time. Cavern of Conventional Science—walls that are dogmas, from which drips ancient wisdom in a patter of slimy opinions—but we have heard a storm of data outside—

Of beings that march in the sky, and of a beacon on the moon—another dark body crosses the sun. Somewhere near Melida there is cannonading, and another stone falls from the sky, at Irkutsk, Siberia; and unknown grain falls from an unknown world, and there are flashes in the sky when the planet Mars is near.

In a farrago of lights and sounds and forms, I feel the presence of possible classifications that may thread a pattern of attempt to find out something. My attention is attracted by a streak of events that is beaded with little star-like points of light. First we shall find out what we can, as to the moon.

In one of the numbers of the *Observatory*, an eminent authority, in some fields of research, is quoted as to the probable distance of the moon. According to his determinations, the moon is 37 miles away. He explains most reasonably:

he is Mr. G. B. Shaw. But by conventional doctrine, the moon is 240,000 miles away. My own idea is that somewhere between determinations by a Shaw and determinations by a Newcomb, we could find many acceptances.

I prefer questionable determinations, myself, or at any rate examinations that end up with questions or considerable latitude. It may be that as to the volcanoes of the moon we can find material for at least a seemingly intelligent question, if no statements are possible as to the size and the distance of the moon. The larger volcanoes of this earth are about three miles in diameter, though the craters of Haleakla, Hawaii, and Aso San, Japan, are seven miles across. But the larger volcanoes of the relatively little moon are said to be sixty miles across, though several are said to be twice that size. And I start off with just about the impression of disproportionality that I should have, if someone should tell me of a pygmy with ears five feet long.

Is there any somewhat good reason for thinking that the volcanic craters of the little moon are larger than, or particularly different in any other way from, the craters of this earth?

If not, we have a direct unit of measurement, according to which the moon is not 2,160, but about 100, miles in diameter.

How far away does one suppose to be an object with something like that diameter, and of the seeming size of the moon?

The astronomers explain. They argue that gravitation must be less powerful upon the moon than upon this earth, and that therefore larger volcanic formations could have been cast up on the moon. We explain. We argue that volcanic force must be less powerful upon the moon than upon this earth, and that therefore larger volcanic formations could not have been cast up on the moon.

The disproportionality that has impressed me has offended more conventional æsthetics than mine. Prof. See, for instance, has tried to explain that the lunar formations are not craters but are effects of bombardment by vast meteors, which spared this earth, for some reason not made clear. Viscid moon—meteor pops in—up splash walls and a central cone. If Prof. See will jump in swimming some day, and then go back some weeks later to see how big a splash he made, he will have other ideas upon such supposed persistences. The moon would have to have been virtually liquid to fit his theory, because there are no partly embedded, vast, round meteors protruding anywhere.

There have been lights like signals upon the moon. There are two conventional explanations: reflected sunlight and volcanic action. Of course, ultra-conventionalists do not admit that in our own times there has been even volcanic action upon the moon. Our instances will be of lights upon the dark part of the moon, and there are good reasons for thinking that our data do not relate to volcanic action. In volcanic eruptions upon this earth the glow is so accompanied by great volumes of smoke that a clear, definite point of light would seem not to be the appearance from a distance.

For Webb's account of a brilliant display of minute dots and streaks of light, in the Mare Crisium, July 4, 1832, see *Astro. Reg.*; 20-165. I have records of half a dozen similar illuminations here, in about 120 years, all of them when the Mare Crisium was in darkness. There can be no commonplace explanation for such spectacles, or they would have occurred oftener; nevertheless the Mare Crisium is a wide, open region, and at times there may have been uncommon percolations of sunlight, and I shall list no more of these interesting events that seem to me to have been like carnivals upon the moon.

Dec. 22, 1835—the star-like light in Aristarchus—reported by Francis Bailey —see Proctor's *Myths and Marvels*, p. 329.

Feb. 13, 1836—in the western crater of Messier—according to Gruithuisen (*Sci. Amer. Sup.*, 7-2629)—two straight lines of light; between them a dark band that was covered with luminous points.

Upon the nights of March 18 and 19, 1847, large luminous spots were seen upon the dark part of the moon, and a general glow upon the upper limb, by the Rev. T. Rankin and Prof. Chevalier (*Rept. B. A.*, 1847-18). The whole shaded part of the disc seemed to be a mixture of lights and shades. Upon the night of the 19th, there was a similar appearance upon this earth, an aurora, according to the London newspapers. It looks as if both the moon and this earth were affected by the same illumination, said to have been auroral. I offer this occurrence as indication that the moon is nearby, if moon and earth could be so affected in common.

But by signaling, I mean something like the appearance that was seen, by Hodgson, upon the dark part of the moon, night of Dec. 11, 1847—a bright light that flashed intermittently. Upon the next night it was seen again (*Monthly Notices* R. A. S., 8-55).

The oppositions of Mars occur once in about two years. and two months. In conventional terms, the eccentricity of the orbit of Mars is greater than the eccentricity of the orbit of this earth, and the part of its orbit that is traversed by this earth in August is nearest the orbit of Mars. When this earth is between Mars and the sun, Mars is said to be in opposition, and this is the position of nearest approach: when opposition occurs in August, that is the most favorable opposition. After that, every two years and about two months, the oppositions are less favorable, until the least favorable of all, in February, after which favorableness increases up to the climacteric opposition in August again. This is a cycle of changing proximities within a period of about fifteen years.

In October, 1862, Lockyer saw a spot like a long train of clouds on Mars, and several days later Secchi saw a spot on Mars. And if that were signaling, it is very meager material upon which to suppose anything. And May 8-22, 1873— white spots on Mars. But, upon June 17, 1873, two months after nearest approach, but still in the period of opposition of Mars, there was either an extraordinary occurrence, or the extraordinariness is in our interpretation. See *Rept. B. A.*, 1874-272. A luminous object came to this earth, and was seen and heard upon the night of June 17, 1873, to explode in the sky of Hungary, Austria, and Bohemia. In the words of various witnesses, termed according to their knowledge, the object was seen seemingly coming from Mars, or from "the red star in the south," where Mars was at the time. Our data were collected by Dr. Galle. The towns of Rybnik and Ratibor, Upper Silesia, are 15 miles apart. Without parallax, this luminous thing was seen from these points "to emerge and separate itself from the disc of the planet Mars." It so happens that we have a definite observation from one of these towns. At Rybnik, Dr. Sage was looking at Mars, at the time. He saw the luminous object "apparently issue from the planet." There is another circumstance, and for its reception our credulity, or our enlightenment, has been prepared. If this thing did come from Mars, it came from the planet to the point where it exploded in about 5 seconds: from the point of explosion, the sound traveled in several minutes. We have a description from Dr. Sage that indicates that a bolt of some kind, perhaps electric, did shoot from Mars, and that the planet quaked with the shock—"Dr. Sage was looking attentively at the planet Mars, when he thus saw the meteor apparently issue from it, and the planet appear as if it was breaking up and dividing into two parts."

Some of the greatest surprises in commonplace experience are discoveries of the nearness of that which was supposed to be the inaccessibly remote.

It seems that the moon is close to this earth, because of the phenomenon of "earthshine." The same appearance has been seen upon the planet Venus. If upon the moon, it is light reflecting from this earth and back to this earth, what is it upon Venus? It is "some unexplained optical illusion" says Newcomb (*Popular Astronomy*, p. 296). For a list of more than twenty observations upon this illumination of Venus, see *Rept. B. A.*, 1873-404. It is our expression that the phenomenon is "unexplained" because it does indicate that Venus is millions of miles closer to this earth than Venus "should" be.

Unknown objects have been seen near Venus. There were more than thirty such observations in the eighteenth century, not relating to so many different periods, however. Our own earliest datum is Webb's observation, of May 22, 1823. I know of only one astronomer who has supposed that these observations could relate to a Venusian satellite, pronouncedly visible sometimes, and then for many years being invisible: something else will have to be thought of. If these observations and others that we shall have, be accepted, they relate to unknown bulks that have, from outer space, gone to Venus, and have been in temporary suspension near the planet, even though the shade of Sir Isaac Newton would curdle at the suggestion. If, acceptably, from outer space, something could go to the planet Venus, one is not especially startled with the idea that something could sail out from the planet Venus—visit this earth, conceivably.

In the *Rept. B. A.*, 1852-8, 35, it is said that, early in the morning of Sept. u, 1852, several persons at Fair Oaks, Staffordshire, had seen, in the eastern sky, a luminous object. It was first seen at 4:15 A.M. It appeared and disappeared several times, until 4:45 A.M., when it became finally invisible. Then, at almost the same place in the sky, Venus was seen, having risen above the eastern horizon. These persons sent the records of their observations to Lord Wrottesley, an astronomer whose observatory was at Wolverhampton. There is published a letter from Lord Wrottesley, who says that at first he had thought that the supposititiously unknown object was Venus, with perhaps an extraordinary halo, but that he had received from one of the observers a diagram giving such a position relatively to the moon that he hesitated so to identify. It was in the period of nearest approach to this earth by Venus, and, since inferior conjunction (July 20, 1852) Venus had been a "morning star." If this thing in the sky were not Venus, the circumstances are that an object came close to this earth, perhaps, and for a while was stationary, as if waiting for the planet Venus to appear above the eastern horizon, then disappearing, whether to sail to Venus or not. We think that perhaps this thing did come close to this earth, because it was, it seems, seen only in the local sky of Fair Oaks. However, if, according to many of our data, professional astronomers have missed extraordinary appearances at reasonable hours, we can't conclude much from what was not reported by them, after 4 o'clock in the morning. I do not know whether this is the origin of the convention or not, but this is the first note I have upon the now standardized explanation that, when a luminous object is seen in the sky at the time of nearest approach by Venus, it is Venus, attracting attention by her great brilliance, exciting persons, unversed in astronomic matters, into thinking that a strange object had visited this earth. When reports are definite as to motions of a seemingly sailing or exploring, luminous thing, astronomers say that it was a fire-balloon.

In the *Rept. B. A.*, 1856-54, it is said that, according to "Mrs. Ayling and friends," in a letter to Lord Wrottesley, a bright object had been seen in the sky of Petworth, Sussex, night of Aug. 11, 1855. According to description, it rose from behind hills, in the distance, at half past eleven o'clock. It was a red body, or it was a red-appearing construction, because from it were projections like spokes of a wheel; or they were "stationary rays," in the words of the description. "Like a red moon, it rose slowly, and diminished slowly, remaining visible one hour and a half." Upon Aug. 11, 1855, Venus was two weeks from primary greatest brilliance, inferior conjunction occurring upon September 30. The thing could not have been Venus, ascending in the sky, at this time of night. An astonishing thing, like a red moon, perhaps with spokes like a wheel's, might, if reported from nowhere else, be considered something that came from outer space so close to this earth that it was visible only in a local sky, except that it might have been visible in other places, and even half past eleven at night may be an unheard-of hour for astronomers, who specialize upon sunspots for a reason that is clearing up to us. Of course an ordinary fire-balloon could be extraordinarily described.

June 8, 1868—I have not the exact time, but one does suspect that it was early in the evening—an object that was reported from Radcliffe Observatory, Oxford. It looked like a comet, but inasmuch as it was reported only from Radcliffe, it may have been in the local sky of Oxford. It seemed to sail in the sky: it moved and changed its course. At first it was stationary; then it moved westward, then southward, then turning north, visible four minutes. See *Eng. Mec.*, 7-351. According to a correspondent to the *Birmingham Gazette*, May 28, 1868, there had been an extraordinary illumination upon Venus, some nights before: a red spot, visible for a few seconds, night of May 27. In the issue of the *Gazette*, of June 1st, someone else writes that he saw this light appearing and disappearing upon Venus. Upon March 15, Browning had seen something that looked like a little shaft of light from Venus (*Eng. Mec.*, 40-130); and upon April 6, Webb had seen a similar appearance (*Celestial Objects*, p. 57). At the time of the appearance at Oxford, Venus was in the period of nearest approach (inferior conjunction July 16, 1868).

I think, myself, that there was one approximately great, wise astronomer. He was Tycho Brahé. For many years, he would not describe what he saw in the sky, because he considered it beneath his dignity to write a book. The undignified, or more or less literary, or sometimes altogether too literary, astronomers, who do write books, uncompromisingly say that when a luminous object is said to have moved to greater degree than could be considered illusory, in a local sky of this earth, it is a fire-balloon. It is not possible to find in the writings of astronomers who so explain, mention of the object that was seen by Coggia, night of Aug. 1, 1871. It seems that this thing was not far away, and did appear only in a local sky of this earth, and if it did come from outer space, how it could have "boarded" this earth, if this earth moves at a rate of 19 miles a second, or 1 mile a second, is so hard to explain that why Proctor and Hind, with their passionate itch for explaining, never took the matter up, I don't know. Upon Aug. 1, 1871, an unknown luminous object was seen in the sky of Marseilles, by Coggia (*Comptes Rendus*, 73-398). According to description, it was a magnificent red object. It appeared at 10:43 P.M., and moved eastward, slowly, until 10:52:30. It stopped—moved northward, and again, at 10:59:30, was stationary. It turned eastward again, and, at 11:3:20, disappeared, or fell behind the horizon. Upon this night Venus was within three weeks of primary greatest brilliance, inferior conjunction occurring upon Sept. 25, 1871.

15

ONE repeating mystery—the mystery of the local sky.

How, if this earth be a moving earth, could anything sail to, fall to, or in any other way reach this earth, without being smashed into fine particles by the impact?

This earth is supposed to rip space at a rate of about 19 miles a second.

Concepts smash when one tries to visualize such an accomplishment.

Now, three times over, we shall have other aspects of this one mystery of the local sky. First we shall take up data upon seeming relation between a region of this earth that is subject to earthquakes, or so-called earthquakes, and appearances in the sky of this especial region, and the repeating falls of objects and substances from this local sky and nowhere else at the times.

We have had records of quakes that occurred at Irkutsk, Siberia, and of stones that fell from the sky to Irkutsk. Upon March 8, 1829, a severe quake, preceded by clattering sounds, was felt at Irkutsk. There was something in the sky. Dr. Erman, the geologist, was in Irkutsk, at the time. In the *Report of the British Association*, 1854-20, it is said that, in Dr. Erman's opinion, the sounds that preceded the quake were in the sky.

The situation at Comrie, Perthshire, is similar. A stone fell, May 17, 1830, in the "earthquake region" around Comrie. It fell at Perth, 22 miles from Comrie. See *Fletcher's List*, p. 100. Upon Feb. 15, 1837, a black powder fell upon the Comrie region (*Edin. New Phil. Jour.*, 31-293). Oct. 12, 1839—a quake at Comrie. According to the Rev. M. Walker, of Comrie, the sky, at the time, was "peculiarly strange and alarming, and appeared as if hung with sackcloth." In Mallet's Catalogue (*Rept. B. A.*, 1854-290) it is said that, throughout the month of October, shocks were felt at Comrie, sometimes slight and sometimes severe—"like distant thunder or reports of artillery"—"the noise sometimes

seemed to be high in the air, and was often heard without any sensible shock."
Upon the 23rd of October, occurred the most violent quake in the whole series
of phenomena at Comrie. See the *Edin. New Phil. Jour.*, vol. 32. All data in
this publication were collected by David Milne. According to the Rev. M.
Maxton, of Foulis Manse, ten miles from Comrie, rattling sounds were heard
in the sky, preceding the shock that was felt. In vol. 33, p. 373, of the Journal,
someone who lived seven miles from Comrie is quoted: "In every case, I am
inclined to say that the sound proceeded not from underground. The sound
seemed high in the air." Someone who lived at Gowrie, forty miles from
Comrie, is quoted: "The most general opinion seems to be that the noise
accompanying the concussion proceeded from above." See vol. 34, p. 87:
another impression of explosion overhead and concussion underneath: "The
noises heard first seemed to be in the air, and the rumbling sound in the earth."
Milne's own conclusion—"It is plain that there are, connected with the
earthquake shocks, sounds both in the earth and in the air, which are distinct
and separate." If, upon the 23rd of October, 1839, there was a tremendous
shock, not of subterranean origin, but from a great explosion in the sky of
Comrie, and if this be accepted, there will be concussions somewhere else.
The "faults" of dogmas will open; there will be seismic phenomena in science.
I have a feeling of a conventional survey of this Scottish sky: vista of a fair,
blue, vacant expanse—our suspicions daub the impression with black alarms
—but also do we project detonating stimulations into the fair and blue, but
unoccupied and meaningless. One cannot pass this single occurrence by,
considering it only in itself: it is one of a long series of quakes of the earth at
Comrie and phenomena in the sky at Comrie. We have stronger evidence than
the mere supposition of many persons, in and near Comrie, that, upon Oct. 23,
1839, something had occurred in the sky, because sounds seemed to come
from the sky. Milne says that clothes, bleaching on the grass, were entirely
covered with black particles which presumably had fallen from the sky. The
shocks were felt in November: in November, according to Milne, a powder
like soot fell from the sky, upon Comrie and surrounding regions. In his report
to the British Association, 1840, Milne, reviewing the phenomena from the
year 1788, says: "Occasionally there was a fall of fine, black powder."

Jan. 8, 1840—sounds like cannonading, at Comrie, and a crackling sound in
the air, according to some of the residents. Whether they were sounds of
quakes or concussions that followed explosions, 247 occurrences, between
Oct. 3, 1839, and Feb. 14, 1841, are listed in the *Edin. New Phil. Jour.*, 32-
107. It looks like bombardment, and like most persistent bombardment—from
somewhere—and the frequent fall from the sky of the débris of explosions.
Feb. 18, 1841a shock and a fall of discolored rain at Comrie (*Edin. New Phil.
Jour.*, 35-148). See Roper's *List of Earthquakes*—year after year, and the
continuance of this seeming bombardment in one small part of the sky of this

earth, though I can find records only of dates and no details. However, I think I have found record of a fall from the sky of débris of an explosion, more substantial than finely powdered soot, at Crieff, which is several miles from Comrie. In the *Amer. Jour. Sci.*, 2-28-275, Prof. Shepard tells a circumstantial story of an object that looked like a lump of slag, or cinders, reported to have fallen at Crieff. Scientists had refused to accept the story, upon the grounds that the substance was not of "true meteoric material." Prof. Shepard went to Crieff and investigated. He gives his opinion that possibly the object did fall from the sky. The story that he tells is that, upon the night of April 23, 1855, a young woman, in the home of Sir William Murray, Achterlyre House, Crieff, saw, or thought she saw, a luminous object falling, and picked it up, dropping it, because it was hot, or because she thought it was hot.

For a description, in a letter, presumably from Sir William Murray, or some member of his family, see *Year Book of Facts*, 1856-273.

It is said that about 12 fragments of scorious matter, hot and emitting a sulphurous odor, had fallen.

In Ponton's *Earthquakes*, p. 118, it is said that, upon the 8th of October, 1857, there had been, in Illinois, an earthquake, preceded by "a luminous appearance, described by some as a meteor and by others as vivid flashes of lightning." Though felt in Illinois, the center of the disturbance was at St. Louis, Mo. One notes the misleading and the obscuring of such wording: in all contemporaneous accounts there is no such indefiniteness as one description by "some" and another notion by "others." Something exploded terrifically in the sky, at St. Louis, and shook the ground "severely" or "violently," at 4:20 A.M., Oct. 8, 1857. According to Timbs' *Year Book of Facts*, 1858-271, "a blinding meteoric ball from the heavens" was seen. "A large and brilliant meteor shot across the heavens" (St. Louis Intelligencer, October 8). Of course the supposed earthquake was concussion from an explosion in the sky, but our own interest is in a series that is similar to others that we have recorded. According to the *New York Times*, October 12, a slight shock was said to have been felt four hours before the great concussion, and another three days before. But see Milne's *Catalog of Destructive Earthquakes*—not a mention of anything that would lead one away from safe and standardized suppositions. See *Bull. Seis. Soc. Amer.*, 3-68—here the "meteor" is mentioned, but there is no mention of the preceding concussions. Time after time, in a period of about three days, concussions were felt in and around St. Louis. One of these concussions, with its "sound like thunder or the roar of artillery" (*New York Times*, October 8) was from an explosion in the sky. If the others were of the same origin—how could detonating meteors so repeat in one small local sky, and nowhere else, if this earth be a moving body? If it be said that only by coincidence did a meteor explode over a region where there had been other quakes, here is the question:

How many times can we accept that explanation as to similar series?

In the *Proceedings of the Society for Psychical Research*, 19-144, a correspondent writes that, in Herefordshire, Sept. 24, 1854, upon a day that was "perfectly still, sky cloudless," he had heard sounds like the discharges of heavy artillery, at intervals of about two minutes, continuing several hours. Again the "mystery of the local sky"—if these sounds did come from the sky. We have no data for thinking that they did.

In the London *Times*, Nov. 9, 1858, a correspondent writes that, in Cardiganshire, Wales, he had, in the autumn of 1855, often heard sounds like the discharges of heavy artillery, two or three reports rapidly, and then an interval of perhaps 20 minutes, also with long intervals, sometimes of days and sometimes of weeks, continuing throughout the winter of 1855-56. Upon the 3rd of November, 1858, he had heard the sounds again, repeatedly, and louder than they had been three years before. In the*Times*, November 12, someone else says that, at Dolgelly, he, too, had heard the "mysterious phenomenon," on the 3rd of November. Someone else—that, upon October 13, he had heard the sounds at Swansea. "The reports, as if of heavy artillery, came from the west, succeeding each other at apparently regular intervals, during the greater part of the afternoon of that day. My impression was that the sounds might have proceeded from practicing at Milford, but I ascertained, the following day, that there had been no firing of any kind there." Correspondent to the *Times*, November 20—that, with little doubt, the sounds were from artillery practice at Milford. He does not mention the investigation as to the sounds of October 13, but says that there had been cannon-firing, upon November 3rd, at Milford. *Times*, December 1—that most of the sounds could be accounted for as sounds of blasting in quarries. *Daily News*, November 16 —that similar sounds had been heard, in 1848, in New Zealand, and were results of volcanic action. *Standard*, November 16—that the "mysterious noise" must have been from Devonport, where a sunken rock had been blown up. So, with at least variety these sounds were explained. But we learn that the series began before October 13. Upon the evening of September 28, in the Dartmoor District, at Crediton, a rumbling sound was heard. It was not supposed to be an earthquake, because no vibration of the ground was felt. It was thought that there had been an explosion of gunpowder. But there had been no such terrestrial explosion. About an hour later another explosive sound was heard. It was like all the other sounds, and in one place was thought to be distant cannonading—terrestrial cannonading. See*Quar. Jour. Geolog. Soc. of London*, vol. 15.

Somewhere near Barisal, Bengal, were occurring just such sounds as the sounds of Cardiganshire, which were like the sounds of Melida. In the *Proc. Asiatic Soc. of Bengal*, November, 1870, are published letters upon the Barisal Guns. One writer says that the sounds were probably booming of the surf. Someone else points out that the sounds, usually described as "explosive," were heard too far inland to be traced to such origin. A clear, calm day, in December, 1871—in *Nature*, 53-197, Mr. G. B. Scott writes that, in Bengal, he had heard "a dull, muffled boom, as if of distant cannon"—single detonations, and then two or three in quicker succession.

In the London *Times*, Jan. 20, 1860, several correspondents write as to a sound "resembling the discharge of a gun high in the air" that was heard near Reading, Berkshire, England, Jan. 17, 1860. See the *Times*, January 24th. To say that a meteor had exploded would, at present, well enough account for this phenomenon.

Sounds like those that were heard in Herefordshire, Sept. 24, 1854, were heard later. In the *English Mechanic*, 100-279, it is said that, upon Nov. 9, 1862, the Rev. T. Webb, the astronomer, of Hardwicke, fifteen miles west of Hereford, heard sounds that he attributed to gunfire at Milford Haven, about 85 miles from Hardwicke. Upon Aug. 1, 1865, Mr. Webb saw flashes upon the horizon, at Hardwicke, and attributed them to gunfire at Tenby, upon occasion of a visit by Prince Arthur. Tenby, too, is about 85 miles from Hardwicke. There were other phenomena in a region centering around Hereford and Worcester. Upon Oct. 6, 1863, there was a disturbance that is now listed as an earthquake; but in the London newspapers so many reports upon this occurrence state that a great explosion had been thought to occur, and that the quake was supposed to be an earthquake of subterranean origin only after no terrestrial explosion could be heard of, that the phenomenon is of questionable origin. There was a similar concussion in about the same region, Oct. 30, 1868. Again the shock was widely attributed to a great explosion, perhaps in London, and again was supposed to have been an earthquake when no terrestrial explosion could be heard of.

Arcana of Science, 1829-196:

That, near Mhow, India, Feb. 27, 1828, fell a stone "perfectly similar" to the stone that fell near Allahabad, in 1802, and a stone that fell near Mooradabad, in 1808. These towns are in the Northwestern Provinces of India.

I have looked at specimens of these stones, and in my view they are similar. They are of brownish rock, streaked and spotted with a darker brown. A stone that fell at Chandakopur, in the same general region, June 6, 1838, is like them. All are as much alike as "erratics" that, because they are alike, geologists ascribe to the same derivation, stationary relatively to the places in which they are found.

It seems acceptable that, upon July 15 and 17, 1822, and then upon a later date, unknown seeds fell from the sky to this earth. If these seeds did come from some other world, there is another mystery as well as that of repetition in a local sky of this earth. How could a volume of seeds remain in one aggregation; how could the seeds be otherwise than scattered from Norway to Patagonia, if they met in space this earth, and if this earth be rushing through space at a rate of 19 miles a second? It may be that the seeds of 1822 fell again. According to Kaemtz (*Meteorology*, p. 465) yellowish brown corpuscles, some round, a few cylindrical, were found upon the ground, June, 1830, near Griesau, Silesia. Kaemtz says that they were tubercules from roots of a well-known Silesian plant—stalk of the plant dries up; heavy rain raises these tubercules to the ground—persons of a low order of mentality think that the things had fallen from the sky. Upon the night of March 24-25, 1852, a great quantity of seeds did fall from the sky, in Prussia, in Heinsberg, Erklenz, and Juliers, according to M. Schwann, of the University of Liége, in a communication to the Belgian Academy of Science (*La Belgique Horticole*, 2-319).

In *Comptes Rendus*, 5-549, is Dr. Wartmann's account of water that fell from the sky, at Geneva. At nine o'clock, morning of Aug. 9, 1837, there were clouds upon the horizon, but the zenith was clear. It is not remarkable that a little rain should fall now and then from a clear sky: we shall see wherein this account is remarkable. Large drops of warm water fell in such abundance that people were driven to shelter. The fall continued several minutes and then stopped. But then, several times during an hour, more of this warm water fell from the sky. *Year Book of Facts*, 1839-262—that upon May 31, 1838, lukewarm water in large drops fell from the sky, at Geneva. *Comptes Rendus*, 15-290—no wind and not a cloud in the sky—at 10 o'clock, morning of May 11, 1842, warm water fell from the sky at Geneva, for about six minutes; five hours later, still no wind and no clouds, again fell warm water, in large drops; falling intermittently for several minutes.

In *Comptes Rendus*, 85-681, is noted a succession of falls of stones in Russia: June 12, 1863, at Buschof, Courland; Aug. 8, 1863, at Pillitsfer, Livonia; April 12, 1864, at Nerft, Courland. Also—see *Fletcher's List*—a stone that fell at Dolgovdi, Volhynia, Russia, June 26, 1864. I have looked at specimens of all four of these stones, and have found them all very much alike, but not of uncommon meteoritic material: all gray stones, but Pillitsfer is darker than the others, and in a polished specimen of Nerft, brownish specks are visible.

In the *Birmingham Daily Post*, June 14, 1858, Dr. C. Mansfield Ingleby, a meteorologist, writes: "During the storm on Saturday (12th) morning, Birmingham was visited by a shower of aerolites. Many hundreds of thousands must have fallen, some of the streets being strewn with them."

Someone else writes that many pounds of the stones had been gathered from awnings, and that they had damaged greenhouses, in the suburbs. In the *Post*, of the 15th, someone else writes that, according to his microscopic examinations, the supposed aerolites were only bits of the Rowley ragstone, with which Birmingham was paved, which had been washed loose by the rain. It is not often that sentiment is brought into meteorology, but in the *Report of the British Association*, 1864-37, Dr. Phipson explains the occurrence meteorologically, and with an unconscious tenderness. He says that the stones did fall from the sky, but that they had been carried in a whirlwind from Rowley, some miles from Birmingham. So we are to sentimentalize over the stones in Rowley that had been torn, by unfeeling paviers, from their companions of geologic ages, and exiled to the pavements of Birmingham, and then some of these little bereft companions, rising in a whirlwind and traveling, unerringly, if not miraculously, to rejoin the exiles. More dark companions. It is said that they were little black stones.

They fell again from the sky, two years later. In *La Science Pour Tous*, June 19, 1860, it is said that, according to the*Wolverhampton Advertiser*, a great number of little black stones had fallen, in a violent storm, at Wolverhampton. According to all records findable by me no such stones have ever fallen anywhere in Great Britain, except at Birmingham and Wolverhampton, which is 13 miles from Birmingham.

Eight years after the second occurrence, they fell again. *English Mechanic*, July 31, 1868—that stones "similar to, if not identical with the well-known Rowley ragstones" had fallen in Birmingham, having probably been carried from Rowley, in a whirlwind.

We were pleased with Dr. Phipson's story, but to tell of more of the little dark companions rising in a whirlwind and going unerringly from Rowley to rejoin the exiles in Birmingham is overdoing. That's not sentiment: that's mawkishness.

In the *Birmingham Daily Post*, May 30, 1868, is published a letter from Thomas Plant, a writer and lecturer upon meteorological subjects. Mr. Plant says, I think, that for one hour, morning of May 29, 1868, stones fell, in Birmingham, from the sky. His words may be interpretable in some other way, but it does not matter: the repeating falls are indication enough of what we're trying to find out—"From nine to ten, meteoric stones fell in immense quantities in various parts of town." "They resembled, in shape, broken pieces of Rowley ragstone ... in every respect they were like the stones that fell in 1858." In the*Post*, June 1, Mr. Plant says that the stones of 1858 did fall from the sky, and were not fragments washed out of the pavement by rain, because many pounds of them had been gathered from a platform that was 20 feet above the ground.

It may be that for days before and after May 29, 1868, occasional stones fell from some unknown region stationary above Birmingham.

In the *Post*, June 2, a correspondent writes that, upon the first of June, his niece, while walking in a field, was struck by a stone that injured her hand severely. He thinks that the stone had been thrown by some unknown person. In the Post, June 4, someone else writes that his wife, while walking down a lane, upon May 24th, had been cut on the head by a stone. He attributes this injury to stone-throwing by boys, but does not say that anyone had been seen to throw the stone.

Symons' Met. Mag., 4-137:

That, according to the *Birmingham Gazette*, a great number of small, black stones had been found in the streets of Wolverhampton, May 25, 1869, after a severe storm. It is said that the stones were precisely like those that had fallen in Birmingham, the year before, and resembled Rowley ragstone outwardly, but had a different appearance when broken.

16

Upon page 287, *Popular Astronomy*, Newcomb says that it is beyond all "moral probability" that unknown worlds should exist in such numbers as have been reported, and should be seen crossing the solar disc only by amateur observers and not by skilled astronomers.

Most of our instances are reports by some of the best-known astronomers.

Newcomb says that for fifty years, prior to his time of writing (edition of 1878) the sun had been studied by such men as Schwabe, Carrington, Secchi, and Spörer, and that they had never seen unknown bodies cross the sun—

Aug. 30, 1863—an unknown body that was seen by Spörer to cross the sun (Webb, *Celestial Objects*, p. 45).

Sept. 1, 1859—two star-like objects that were seen by Carrington to cross the sun (*Monthly Notices*, 20-13, 15, 88).

Things that crossed the sun, July 31, 1826, and May 26, 1828— see *Comptes Rendus*, 83-623, and Webb's *Celestial Objects*, p. 40. From Sept. 6 to Nov. 1, 1831, an unknown luminous object was seen every cloudless night, at Geneva, by Dr. Wartmann and his assistants (*Comptes Rendus*, 2-307). It was reported from nowhere else. What all the other astronomers were doing, September-

October, 1831, is one of the mysteries that we shall not solve. An unknown, luminous object that was seen, from May 11 to May 14, 1835, by Cacciatore, the Sicilian astronomer (*Amer. Jour. Sci.*, 31-158). Two unknowns that, according to Pastorff, crossed the sun, Nov. 1, 1836, and Feb. 16, 1837 (*An. Sci. Disc.*, 1860-410)—De Vico's unknown, July 12, 1837 (*Observatory*, 2-424)-observation by De Cuppis, Oct. 2, 1839 (*C. R.*, 83-314)-by Scott and Wray, last of June, 1847; by Schmidt, Oct. 11, 1847 (*C. R.*, 83-623)-two dark bodies that were seen, Feb. 5, 1849, by Brown, of Deal (*Rec. Sci.*, 1-138)—object watched by Sidebotham, half an hour, March 12, 1849, crossing the sun (*C. R.*, 83-622)—Schmidt's unknown, Oct. 14, 1849 (*Observatory*, 3-137)—and an object that was watched, four nights in October, 1850, by James Ferguson, of the Washington Observatory. Mr. Hind believed this object to be a Trans-Neptunian planet, and calculated for it a period of 1,600 years. Mr. Hind was a great astronomer, and he miscalculated magnificently: this floating island of space was not seen again (*Smithson. Miscell. Cols.*, 20-20).

About May 30, 1853—a black point that was seen against the sun, by Jaennicke (*Cosmos*, 20-64).

A procession—in the *Rept. B. A.*, 1855-94, R. P. Greg says that, upon May 22, 1854, a friend of his saw, near Mercury, an object equal in size to the planet itself, and behind it an elongated object, and behind that something else, smaller and round.

June 11, 1855—a dark body of such size that it was seen, without telescopes, by Ritter and Schmidt, crossing the sun (*Observatory*, 3-137). Sept. 12, 1857 —Ohrt's unknown world; seemed to be about the size of Mercury (*C. R.*, 83-623)—Aug. 1, 1858-unknown world reported by Wilson, of Manchester (*Astro. Reg.*, 9-287).

I am not listing all the unknowns of a period; perhaps the object reported by John H. Tice, of St. Louis, Mo., Sept. 15, 1859, should not be included; Mr. Tice was said not to be trustworthy—but who has any way of knowing? However, I am listing enough of these observations to make me feel like a translated European of some centuries ago, relatively to a wider existence— lands that may be the San Salvadors, Greenlands, Madagascars, Cubas, Australias of extra-geography, all of them said to have crossed the sun, whereas the sun may have moved behind some of them

Jan. 29, 1860—unknown object, of planetary size, reported from London, by Russell and three other observers (*Nature*, 15-505). Summer of 1860—see *Sci. Amer.*, 35-340, for an account, by Richard Covington, of an object that, without a telescope, he saw crossing the sun. An unknown world, reported by Loomis, of Manchester, March 20, 1862 (*Monthly Notices*, 22-232)—a newspaper account of an object that was seen crossing the sun, Feb. 12, 1864, by Samuel Beswick, of New York (*Astro. Reg.*, 2-161)—unknown that was seen, March 18, 1865, at Constantinople (*L'Ann. Sci.*, 1865-16)—unknown "cometic objects" that were seen, Nov. 4, 9, and 18, 1865 (*Monthly Notices*, 26-242).

Most of these unknowns were seen in the daytime. Several reflections arise. How could there be stationary regions over Irkutsk, Comrie, and Birmingham, and never obscure the stars—or never be seen to obscure the stars? A heresy that seems too radical for me is that they may be beyond the nearby stars. A more reasonable idea is that if nightwatchmen and policemen and other persons who do stay awake nights, should be given telescopes, something might be found out. Something else that one thinks of is that, if so many unknowns have been seen crossing the sun, or crossed by the sun, others not so revealed must exist in great numbers, and that instead of being virtually blank, space must be archipelagic.

Something that was seen at night; observer not an astronomer—

Nov. 6, 1866—an account, in the London *Times*, Jan. 2, 1867, by Senor De Fonblanque, of the British Consulate, at Cartagena, U. S. Colombia, of a luminous object that moved in the sky. "It was of the magnitude, color, and brilliance of a ship's red light, as seen at a distance of 200 yards." The object was visible three minutes, and then disappeared behind buildings. De Fonblanque went to an open space to look for it, but did not see it again.

17

IF we could stop to sing, instead of everlastingly noting vol. this and p. that, we could have the material of sagas—of the bathers in the sun, which may be neither intolerably hot nor too uncomfortably cold; and of the hermit who floats across the moon; of heroes and the hairy monsters of the sky. I should stand in public places and sing our data—sagas of parades and explorations and massacres in the sky—having a busy band of accompanists, who set off fireworks, and send up balloons, and fire off explosives at regular intervals— extra-geographic songs of boiling lakes and floating islands—extra-sociologic meters that express the tramp of space-armies upon inter-planetary paths covered with little black pebbles—biologic epics of the clouds of mammoths and horses and antelopes that once upon a time fell from the sky upon the northern coast of Siberia—

Song that interprets the perpendicular white streaks in the repeating mirages at Youghal—the rhythmic walruses of space that hang on by their tusks to the edges of space-islands, sometimes making stars variable as they swing in cosmic undulations—so a round space-island with its border of gleaming tusks, and we frighten children with the song of an ogre's head, with a wide-open mouth all around it—fairy lands of the little moon, and the tiny civilizations in rocky cups that are sometimes drained to their slums by the wide-mouthed ogres. The Maelstrom of Everlasting Catastrophe that overhangs Genoa, Italy—and twines its currents around a living island. The ground underneath quakes with the struggle—then the fall of blood—and the fall of blood—three days the fall of blood from the broken red brooks of a living island whose mutilations are scenery—

But after all, it may be better that we go back to *Rept. B. A.*—see vol. 1849, p. 46—a stream of black objects, crossing the sun, watched, at Naples, May II, 1845, by Capocci and other astronomers—things that may have been seeds.

A great number of red points in the sky of Urrugne, July 9, 1853 (*An. Soc. Met. de France*, 1853-227). *Astro. Reg.*, 5-179—C. L. Prince, of Uckfield, writes that, upon June 11, 1867, he saw objects crossing the field of his telescope. They were seeds, in his opinion.

Birmingham Daily Post, May 31, 1867:

Mr. Bird, the astronomer, writes that, about 11 A.M., May 30, he saw unknown forms in the sky. In his telescope, which was focused upon them and upon the planet Venus, they appeared to be twice the size of Venus. They were far away, according to focus; also, it may be accepted that they were far away because an occasional cloud passed between them and this earth. They did not move like objects carried in the wind: all did not move in the same direction, and they moved at different speeds.

"All of them seemed to have hairy appendages, and in many cases a distinct tail followed the object and was highly luminous." Flashes that have been seen in the sky—and they're from a living island that wags his luminous peninsula. Hair-like substances that have fallen to this earth—a meadow has been shorn from a monster's mane. My animation is the notion that it is better to think in tentative hysteria of pairs of vast things, traveling like a North and South America through the sky, perhaps one biting the other with its Gulf of Mexico, than to go on thinking that all things that so move in the sky are seeds, whereas all things that swim in the sea are not sardines.

In the *Post*, June 3, 1867, Mr. W. H. Wood writes that the objects were probably seeds. *Post*, June 5—Mr. Bird says that the objects were not seeds. "My intention was simply to describe what was seen, and the appearance was certainly that of meteors." He saves himself, in the annals of extra-geography —"whether they were meteors of the ordinary acceptation, is another matter."

And the planet Venus, and her veil that is dotted with blue-fringed cupids—in the *Astronomical Register*, 7-138, a correspondent writes, from Northampton, that, upon May 2, 1869, he was looking at Venus, and saw a host of shining objects, not uniform in size. He thinks that it is unlikely that so early in the spring could these objects be seeds. He watched them about an hour and twenty minutes—"many of the larger ones were fringed on one side, the fringe appearing somewhat bluish." Or that it is better even to sentimentalize than to go on stupidly thinking that all such things in the sky are seeds, whereas all things in the sea are not the economically adjusting little forms without which critics of underground traffic in New York probably could not express themselves—the planet Venus—she approaches this lordly earth—the blue-fringed ecstasies that suffuse her skies.

With the phenomena of Aug. 7, 1869, I suspect that the "phantom soldiers" that have been seen in the sky, may have been reflections from, or mirages of, things or beings that march, in military formations, in space. In *Popular Astronomy*, 3-159, Prof. Swift writes that, at Mattoon, Ill., during the eclipse of the sun, of Aug. 7, 1869, he had seen, crossing the moon, objects that he thought were seeds. If they were seeds, also there happened to be seeds in the sky of Ottumwa, Iowa: here, crossing the visible part of the sun, twenty minutes before totality of the eclipse, Prof. Himes and Prof. Zentmayer saw objects that marched, or that moved, in straight, parallel lines (*Les Mondes*, 21-241). In the *Jour. Frank. Inst.*, 3-58-214, it is said that some of these objects moved in one direction across the moon, and that others moved in another direction across another part of the moon, each division moving in parallel lines. If these things were seeds, also there happened to be seeds in the sky, at Shelleyville, Kentucky. Here were seen, by Prof. Winlock, Alvan Clark, Jr., and George W. Dean, things that moved across the moon, during the eclipse, in parallel, straight lines (*Pop. Astro.*, 2-332).

Whatever these things may have been, I offer another datum indicating that the moon is nearby: that these objects probably were not, by coincidence, things in three widely separated skies, parallelness giving them identity in two of the observations; and, if seen, without parallax, from places so far apart, against the moon, were close to the moon; that observation of such detail would be unlikely if they were near a satellite 240,000 miles away—unless, of course, they were mountain-sized.

It may be that out from two floating islands of space, two processions had marched across the moon. *Observatory*, 3-137—that, at St. Paul's Junction, Iowa, four persons had seen, without telescopes, a shining object close to the sun and moon, apparently; that, with a telescope, another person had seen another large object, crescentically illumined, farther from the sun and moon in eclipse. See *Nature*, 18-663, and *Astro. Reg.,*. 7-227.

I have many data upon the fall of organic matter from the sky. Because of my familiarity with many records, it seems no more incredible that up in the seemingly unoccupied sky there should be hosts of living things than that the seeming blank of the ocean should swarm with life. I have many notes upon a phosphorescence, or electric condition of things that fall from the sky, for instance the highly luminous stones of Dhurmsulla, which were intensely cold —*Amer. Jour. Sci.*, 2-28-270:

It is said that, according to investigations by Prof. Shepard, a luminous substance was seen falling slowly, by Sparkman R. Striven, a young man of seventeen, at his home, in Charleston, S. C., Nov. 16, 1857. It is said that the young man saw a fiery, red ball, the size and shape of an orange, strike a fence, breaking, and disappearing. Where this object had struck the fence, was found "a small bristling mass of black fibers." According to Prof. Shepard, it was "a confused aggregate of short clippings of the finest black hair, varying in length from one tenth to one third of an inch." Prof. Shepard says that this substance was not organic. It seems to me that he said this only because of the coercions of his era. My reason for so thinking is that he wrote that when he analyzed these hairs they burned away, leaving grayish skeletons, and that they were "composed in part of carbon," and burned with an odor "most nearly bituminous."

For full details of the following circumstances, see *Comptes Rendus*, 13-215, and *Rept. B. A.*, 1854-302:

Feb. 17, 1841—the fall, at Genoa, Italy, of a red substance from the sky— another fall upon the 18th—a slight quake, at 5 P.M., February 18th—another quake, six hours later—fall of more of the red substance, upon the 19th. Some of this substance was collected and analyzed by M. Canobbia, of Genoa. He says it was oily and red.

18

In a pamphlet entitled Wonderful Phenomena, by Curtis Eli, is the report of an occurrence, or of an alleged occurrence, that was investigated by Mr. Addison A. Sawin, a spiritualist. He interpreted in the only way that I know of, and that is the psychochemic process of combining new data with preconceptions with which they seem to have affinity. It is said that, at Warwick, C. W., Oct. 3, 1843, somebody named Charles Cooper heard a rumbling sound in the sky, and saw a cloud, under which were three human forms, "perfectly white," sailing through the air above him, not higher than the tree-tops. It is said that the beings were angels. They were male angels. That is orthodox. The angels wafted through the air, but without motions of their own, and an interesting observation is that they seemed to have belts around their bodies—as if they had been let down from a vessel above, though this poor notion is not suggested in the pamphlet. They "moaned." Cooper called to some men who were laboring in another field, and they saw the cloud, but did not see the forms of living beings under it. It is said that a boy had seen the beings in the air, "side by side, making a loud and mournful noise." Another person, who lived six miles away, is quoted: "he saw the clouds and the persons and heard the sounds." Mr. Sawin quotes others, who had seen "a remarkable cloud," and had heard the sounds, but had not seen the angels. He ends up: "Yours is the glorious hope of the resurrection of the soul." The gloriousness of it is an inverse function of the dolefulness of it: Sunday Schools will not take kindly to the doctrine—be good and you will moan forever. One supposes that the glorious hope colored the whole investigation.

Some day I shall publish data that lead me to suspect that many appearances upon this earth that were once upon a time interpreted by theologians and demonologists, but are now supposed to be the subject-matter of psychic research, were beings and objects that visited this earth, not from a spiritual existence, but from outer space. That extra-geographic conditions may be spiritual, or of highly attenuated matter, is not my present notion, though that, too, may be some day accepted. Of course all these data suffer, in one way, about as much distortion as they would in other ways, if they had been reported by astronomers or meteorologists. As to all the material in this chapter, I take the position that perhaps there were appearances in the sky, and perhaps they were revelations of, or mirages from, unknown regions and conditions of outer space, and spectacles of relatively nearby inhabited lands, and of space-travelers, but that all reports upon them were products of the assimilating of the unknown with figures and figments of the nearest familiar similarities. Another position of mine that will be found well-taken is that, no matter what my own interpretations or acceptances may be, they will compare favorably, so far as rationality is concerned, with orthodox explanations. There have been many assertions that "phantom soldiers" have been seen in the sky. For the orthodox explanation of the physicists, see Brewster's *Natural Magic*, p. 125: a review of the phenomenon of June 23, 1744; that, according to 27 witnesses, some of whom gave sworn testimony before a magistrate, whether that should be mentioned or not, troops of aërial soldiers had been seen, in Scotland, on and over a mountain, remaining visible two hours and then disappearing because of darkness. In Clarke's *Survey of the Lakes* (fol. 1789) is an account in the words of one of the witnesses. See *Notes and Queries*, 1-7-304. Brewster says that the scene must have been a mirage of British troops, who, in anticipation of the rebellion of 1745, were secretly maneuvering upon the other side of the mountain. With a talent for clear-seeing, for which we are notable, except when it comes to some of our own explanations, we almost instantly recognize that, to keep a secret from persons living upon one side of a mountain, it is a very sensible idea to go and maneuver upon the other side of the mountain; but then how to keep the secret, in a thickly populated country like Scotland, from persons living upon that other side of the mountain—however, there never has been an explanation that did not itself have to be explained.

Or the "phantom soldiers" that were seen at Ujest, Silesia, in 1785—see Parish's *Hallucinations and Illusions*, p. 309. Parish finds that at the time of this spectacle, there were soldiers, of this earth, marching near Ujest; so he explains that the "phantom soldiers" were mirages of them. They were marching in the funeral procession of General von Cosel. But some time later they were seen again, at Ujest—and the General had been dead and buried several days, and his funeral procession disbanded—and if a refraction can survive independently of its primary, so may a shadow, and anybody may take a walk where he went a week before, and see some of his shadows still wandering around without him. The great neglect of these explainers is in not accounting for an astonishing preference for, or specialization in, marching soldiers, by mirages. But if often there be, in the sky, things or beings that move in parallel lines, and, if their betrayals be not mirages, but their shadows cast down upon the haze of this earth, or Brocken specters, such frequency, or seeming specialization, might be accounted for.

Sept. 27, 1846—a city in the sky of Liverpool (*Rept. B. A.*, 1847-39) . The apparition is said to have been a mirage of the city of Edinburgh. This "identification" seems to have been the product of suggestion: at the time a panorama of Edinburgh was upon exhibition in Liverpool.

Summer of 1847—see Flammarion's *The Atmosphere*, p. 160—story told by M. Grellois: that he was traveling between Ghelma and Bône, when he saw, to the east of Bône, upon a gently sloping hill, "a vast and beautiful city, adorned with monuments, domes, and steeples." There was no resemblance to any city known to M. Grellois.

In the *Bull. Soc. Astro. de France*, 21-180, is an account of a spectacle that, according to 20 witnesses, vas seen for two hours in the sky of Vienne dans le Dauphiné, May 3, 1848. A city—and an army, in the sky. One supposes that a Brewster would say that nearby was a terrestrial city, with troops maneuvering near it. But also vast lions were seen in the sky—and that is enough to discourage any Brewster. Four months later, according to the London *Times*, Sept. 13, 1848, a still more discouraging—or perhaps stimulating—spectacle was, or was not, seen in Scotland. Afternoon of Sept. 9, 1848—Quigley's Point, Lough Foyle, Scotland—the sky turned dark. It seemed to open. The opening looked reddish, and in the reddish area, appeared a regiment of soldiers. Then came appearances that looked like war vessels under full sail, then "a man and a woman and a swan and a peahen." The "opening" closed, and that was the last of this shocking or ridiculous mixture that nobody but myself would record as being worth thinking about.

"Phantom soldiers" that were seen in the sky, near the Banmouth, Dec. 30, 1850 (*Rept. B. A.*, 1852-30).

"Phantom soldiers" that were seen at Buderich, Jan. 22, 1854 (*Notes and Queries*, 1-9-267).

"Phantom soldiers" that were seen by Lord Roberts (*Forty-One Years in India*, p. 219) at Mohan, Feb. 25, 1858. It is either that Lord Roberts saw indistinctly, and described in terms of the familiar to him, or that we are set back in our own motions. According to him, the figures wore Hindoo costumes.

Extra-geography—its vistas and openings and fields—and the Thoreaus that are upon this earth, but undeveloped, because they cannot find their ponds. A lonely thing and its pond, afloat in space—they crossed the moon. In *Cosmos*, n. s., 11-200, it is said that, night of July 7, 1857, two persons of Chambon had seen forms crossing the moon—something like a human being followed by a pond.

"Phantom soldiers" that were seen, about the year 1860, at Paderborn, Westphalia (Crowe, *Night-side of Nature*, p. 416).

19

We attempt to co-ordinate various streaks of data, all of which signify to us that, external to this earth, and in relation with, or relatable to, this earth are lands and lives and a generality of conditions that make of the whole, supposed solar system one globule of circumstances like terrestrial circumstances. Our expressions are in physical terms, though in outer space there may be phenomena known as psychic phenomena, because of the solid substances and objects that have fallen from the sky to this earth, similar to, but sometimes not identified with, known objects and substances upon this earth. Opposing us is the more or less well-established conventional doctrine that has spun like a cocoon around mind upon this earth, shutting off research, and stifling even speculation, shelling away all data of relations and relatability with external existences, a doctrine that, in its various explanations and disregards and denials, is unified in one expression of Exclusionism.

An unknown vegetable substance falls from the sky. The datum is buried: it may sprout some day.

The earth quakes. A luminous object is seen in the sky. Substance falls from the unknown. But the event is catalogued with subterranean earthquakes.

All conventional explanations and all conventional disregards and denials have Exclusionism in common. The unity is so marked, all writings in the past are so definitely in agreement, that I now think of a general era that is, by Exclusionism, as distinctly characterized as ever was the Carboniferous Era.

A pregnant woman stands near Niagara Falls. There are sounds, and they are vast circumstances; but the cells of an unborn being respond, or vibrate, only as they do to disturbances in their own little environment. Horizons pour into a gulf, and thunder rolls upward: embryonic consciousness is no more than to slight perturbations of maternal indigestion. It is Exclusionism.

Stones fall from the sky. To the same part of this earth, they fall again. They fall again. They fall from some region that, relatively to this part of the earth's surface, is stationary. But to say this leads to the suspicion that it is this earth that is stationary. To think that is to beat against the walls of uterine dogmas— into a partly hairy and somewhat reptilian mass of social undevelopment comes exclusionist explanation suitable for such immaturity.

It does not matter which of our subjects we take up, our experience is unvarying: the standardized explanation will be Exclusionism. As to many appearances in the sky, the way of excluding foreign forces is to say that they are auroras, which are supposed to be mundane phenomena. School children are taught that auroras are electric manifestations encircling the poles of this earth. Respectful urchins are shown an ikon by which an electrified sphere does have the polar encirclements that it should have. But I have taken a disrespectful, or advanced, course through the *Monthly Weather Review*, and have read hundreds of times of auroras that were not such polar crownings: of auroras in Venezuela, Sandwich Islands, Cuba, India; of an aurora in Pennsylvania, for instance, and not a sign of it north of Pennsylvania. There are lights in the sky for which "auroral" is as good a name as any that can be thought of, but there are others for which some other names will have to be thought of. There have been lights like luminous surfs beating upon the coasts of this earth's atmosphere, and lights like vast reflections from distant fires; steady pencils of light and pulsating clouds and quick flashes and seeming objects with definite outlines, all in one poverty of nomenclature, for which science is, in some respects, not notable, called "auroral." Nobody knows what an aurora is. It does not matter. An unknown light in the sky is said to be auroral. This is standardization, and the essence of this standardization is Exclusionism.

I see one resolute, unified, unscrupulous exclusion from science of the indications of nearby lands in the sky. It may not be unscrupulousness: it may be hypnosis. I see that all seeming hypnotics, or somnambulists of the past, who have most plausibly so explained, or so denied, have prospered and have had renown. According to my impressions, if a Brewster, or a Swift, or a Newcomb ever had written that there may be nearby lands and living beings in the sky, he would not have prospered, and his renown would be still subject to delay. If an organism flourishes, it is said to be in harmony with environment, or with higher forces. I now conceive of successful and flourishing Exclusionism as an organization that has been in harmony with higher forces. Suppose we accept that all general delusions function sociologically. Then, if Exclusionism be general delusion; if we shall accept that conceivably the isolation of this earth has been a necessary factor in the development of the whole geo-system, we see that exclusionistic science has faithfully, though falsely, functioned. It would be world-wide crime to spread world-wide too soon the idea that there are other existences nearby and that they have been seen and that sounds from them have been heard: the peoples of this earth must organize themselves before conceiving of, and trying to establish, foreign relations. A premature science of such subjects would be like a United States taking part in a Franco-Prussian War, when such foreign relations should be still far in the future of a nation that has still to concentrate upon its own internal development.

So in the development of all things—or that a stickleback may build a nest, and so may vaguely and not usefully and not explicably at all, in terms of Darwinian evolution, foreshadow a character of coming forms of life; but that a fish that should try to climb a tree and sing to its mate before even the pterodactyl had flapped around with wings daubed with clay would be an unnoticed little clown in cosmic drama. But I do conceive that when the Carboniferous Era is dominant, and when not a discordant thing will be permitted to flourish, though it may adumbrate, restrictions will not last forever, and that the rich and bountiful curse upon rooted things will some day be lifted.

20

PATCHED by a blue inundation that had never been seen before—this earth, early in the 60's of the 19th century. Then faintly, from far away, this new appearance is seen to be enveloped with volumes of gray. Flashes like lightning, and faintest of rumbling sounds—then cloud-like envelopments roll away, and a blue formation shines in the sun. Meteorologists upon the moon take notes.

But year after year there are appearances, as seen from the moon, that are so characterized that they may not be meteorologic phenomena upon this earth: changing compositions wrought with elements of blue and of gray; it is like conflict between Synthesis and Dissolution: straight lines that fade into scrawls, but that reform into seeming moving symbols: circles and squares and triangles abound.

Having had no mean experience with interpretations as products of desires, given that upon the moon communication with this earth should be desired, it seems likely to me that the struggles of hosts of Americans, early in the 60's of the 19th century, were thought by some lunarians to be maneuvers directed to them, or attempts to attract their attention. However, having had many impressions upon the resistance that new delusions encounter, so that, at least upon this earth, some benightments have had to wait centuries before finally imposing themselves generally, I'd think of considerable time elapsing before the coming of a general conviction upon the moon that, by means of living symbols, and the firing of explosives, terrestrians were trying to communicate.

Beacon-like lights that have been seen upon the moon. The lights have been desultory. The latest of which I have record was back in the year 1847. But now, if beginning in the early 60's, though not coinciding with the beginning of unusual and tremendous manifestations upon this earth, we have data as if of greatly stimulated attempts to communicate from the moon—why one assimilates one's impressions of such great increase with this or with that, all according to what one's dominant thoughts may be, and calls the product a logical conclusion. Upon the night of May 15, 1864, Herbert Ingall, of Camberwell, saw a little to the west of the lunar crater Picard, in the Mare Crisium, a remarkably bright spot (*Astro. Reg.*, 2-264).

Oct. 24, 1864—period of nearest approach by Mars—red lights upon opposite parts of Mars (*C. R.*, 85-538). Upon Oct. 16, Ingall had again seen the light west of Picard. Jan. I, 1865—a small speck of light, in darkness, under the east foot of the lunar Alps, shining like a small star, watched half an hour by Charles Grover (*Astro. Reg.*, 3-255). Jan. 3, .1865—again the red lights of Mars (*C. R.*, 85-538). A thread of data appears, as an offshoot from a main streak, but it cannot sustain itself. Lights on the moon and lights on Mars, but I have nothing more that seems to signify both signals and responses between these two worlds.

April 10, 1865—west of Picard, according to Ingall—"a most minute point of light, glittering like a star" (*Astro. Reg.*, 3-189).

Sept. 5, 1865—a conspicuous bright spot west of Picard (*Astro. Reg.*, 3-252).

It was seen again by Ingall. He saw it again upon the 7th, but upon the 8th it had gone, and there was a cloudlike effect where the light had been.

Nov. 24, 1865—a speck of light that was seen by the Rev. W. O. Williams, shining like a small star in the lunar crater Carlini (*Intel. Obs.*, II-58).

June to, 1866—the star-like light in Aristarchus; reported by Tempel (Denning, *Telescopic Work*, p. 121).

Astronomically and seleno-meteorologically, nothing that I know of has ever been done with these data. I think well of taking up the subject theologically. We are approaching accounts of a different kind of changes upon the moon. There will be data seeming so to indicate not only persistence but devotedness upon the moon that I incline to think not only of devotedness but of devotions. Upon the 16th of October, 1866, the astronomer Schmidt, of the land of Socrates, announced that the isolated object, in the eastern part of the Mare Serenitatis, known as Linné, had changed Linné stands out in a blank area like the Pyramid of Cheops in its desert. If changes did occur upon Linné, the conspicuous position. seems to indicate selection. Before October, 1866, Linné was well-known as a dark object. Something was whitening an object that had been black.

A hitherto unpublished episode in the history of theologies:

The new prophet who had appeared upon the moon—

Faint perceptions of moving formations, often almost rigorously geometric, upon one part of this earth, and perhaps faintest of signal-like sounds that reached the moon—the new prophet—and that he preached the old lunar doctrine that there is no god but the Earth-god, but exhorted his hearers to forsake their altars upon which had burned unheeded lights, and to build a temple upon which might be recited a litany of lights and shades.

We are only now realizing how the Earth-god looks to the beings of the moon —who know that this earth is dominant; who see it frilled with the loops of the major planets; its Elizabethan ruff wrought by the complications of the asteroids; the busy little sun that brushes off the dark. God of the moon, when mists make it expressionless—a vast, bland, silvery Buddha.

God of the moon, when seeing is clear—when the disguise is off—when, at night, from pointed white peaks drip the fluctuating red lights of a volcano, this earth is the appalling god of carnivorousness.

Sometimes the great roundish earth, with the heavens behind it broken by refraction, looks like something thrust into a shell from external existence— clouds of tornadoes as if in its grasp—and it looks like the fist of God, clutching rags of ultimate fire and confusion.

That a new prophet had appeared upon the moon, and had excited new hope of evoking response from the bland and shining Stupidity that has so often been mistaken for God, or from the Appalling that is so identified with Divinity—from the clutched and menacing fist that has so often been worshiped.

There is no intelligence except era-intelligence. Suppose the whole geo-system be a super-embryonic thing. Then, by the law of the embryo, its parts cannot organize until comes scheduled time. So there are local congeries of development of a chick in an egg, but these local centers cannot more than faintly sketch out relations with one another, until comes the time when they may definitely integrate. Suppose that far back in the 19th century there were attempts to communicate from the moon; but suppose that they were premature: then we suppose the fate of the protoplasmic threads that feel out too soon from one part of an egg to another. In October, 1866, Schmidt, of Athens, saw and reported in terms of the concepts of his era, and described in conventional selenographic language. See *Rept. B. A.*, 1867.

Upon Dec. 14, 16, 25, 27, 1866, Linné was seen as a white spot. But there was something that had the seeming more of a design, or of a pattern, an elaboration upon the mere turning to white of something that had been black—a fine, black spot upon Linné; by Schmidt and Buckingham, in December, 1866 (*The Student*, 1-261). The most important consideration of all is reviewed by Schmidt in the *Rept. B. A.*, 1867-22—that sunlight and changes of sunlight had nothing to do with the changing appearances of Linné.

Jan. 14, 1867—the white covering, or, at least, seeming of covering, of Linné, had seemingly disappeared—Knott's impression of Linné as a dark spot, but "definition" was poor. January 16—Knott's very strong impression, which, however, he says may have been an illusion, of a small central dark spot upon Linné. Dawes' observation, of March 15, 1867—"an excessively minute black dot in the middle of Linné."

A geometric figure that was white-bordered and centered with black, formed and dissolved and formed again.

I have an impression of spectacles that were common in the United States, during the War: hosts of persons arranging themselves in living patterns: flags, crosses, and in one instance, in which thousands were engaged, in the representation of an enormous Liberty Bell. Astronomers have thought of trying to communicate with Mars or the moon by means of great geometric constructions placed conspicuously, but there is nothing so attractive to attention as change, and a formation that could appear and disappear would enchance the geometric with the dynamic. That the units of the changing compositions that covered Linné were the lunarians themselves—that Linné was terraced—hosts of the inhabitants of the moon standing upon the ridges of their Cheops of the Serene Sea, some of them dressed in white and standing in a border, and some of them dressed in black, centering upon the apex, or the dark material of the apex left clear for the contrast, all of them unified in a hope of conveying an impression of the geometric, as the product of design, and distinguishable from the topographic, to the shining god that makes the stars of their heavens marginal.

It is a period of great activity—or of conflicting ideas and purposes—upon the moon: new and experimental demonstrations, but also, of course, the persistence of the old. In the *Astronomical Register*, 5-114, Thomas G. Elger writes that upon the 9th of April, 1867, he was surprised to see, upon the dark part of the moon, a light like a star of the 7th magnitude, at 7:30 P.M. It became fainter, and looked almost extinguished at 9 o'clock. Mr. Elger had seen lights upon the moon before, but never before a light so clear—"too bright to be overlooked by the most careless observer." May 7, 1867—the beacon-like light of Aristarchus—observed by Tempel, of Marseilles, when Aristarchus was upon the dark part of the moon (*Astro. Reg.*, 5-220). Upon the night of June 10, 1867, Dawes saw three distinct, roundish, black spots near Sulpicius Gallus, which is near Linné; when looked for upon the 13th, they had disappeared (*The Student*, 1-261).

Aug. 6, 1867—

And this earth in the sky of the moon—smooth and bland and featureless earth —or one of the scenes that make it divine and appalling—jaws of this earth, as seem to be rims of more or less parallel mountain ranges, still shining in sunlight, but surrounded by darkness—

And, upon the moon, the assembling of the Chiaroscuroans, or the lunar communicationists who seek to be intelligible to this earth by means of lights and shades, patterned upon Linné by their own forms and costumes. The Great Pyramid of Linné, at night upon the moon—it stands out as a bold black triangularity pointing to this earth. It slowly suffuses white—the upward drift of white-clad forms, upon the slopes of the Pyramid. The jaws of this earth seem to munch, in variable light. There is no other response. Devotions are the food of the gods.

Upon Aug. 6, 1867, Buckingham saw upon Linné, which was in darkness, "a rising oval spot" (*Rept. B. A.*, 1867-7). In October, 1867, Linné was seen as a convex white spot (*Rept. B. A.*, 1867-8) .

Also it may be that the moon is not inhabited, and is not habitable. There are many astronomers who say that the moon has virtually no atmosphere, because when a star is passed over by the moon, the star is not refracted, according to them. See Clerke's *History of Astronomy*, p. 264—that, basing his calculations upon the fact that a star is never refracted out of place when occulted by the moon, Prof. Comstock, of Washburn Observatory, had determined that this earth's atmosphere is 5,000 times as dense as the moon's.

I did think that in this secondary survey of ours we had pretty well shaken off our old opposition, the astronomers: however, with something of the kindliness that one feels for renewed meeting with the familiar, here we are at home with the same old kind of demonstrations: the basing of laborious calculations upon something that is not so

See index of *Monthly Notices, R. A. S.*—many instances of stars that have been refracted out of place when occulted by the moon. See the *Observatory*, 24-210, 313, 315, 345, 414; *English Mechanic*, 23-197, 279; 26-229; 52—index, "atmosphere"; 81-60; 84-161; 85-108.

In the year 1821, Gruithuisen announced that he had discovered a city of the moon. He described its main thoroughfare and branching streets. In 1826, he announced that there had been considerable building, and that he had seen new streets. This formation, which is north of the crater Schroeter, has often been examined by disagreeing astronomers: for a sketch of it, in which a central line and radiating lines are shown, see the *English Mechanic*, 18-638. There is one especial object upon the moon that has been described and photographed and sketched so often that I shall not go into the subject. For many records of observations, see the *English Mechanic* and *L'Astronomie*. It is an object shaped like a sword, near the crater Birt. Anyone with an impression of the transept of a cathedral, may see the architectural here. Or it may be a mound similar to the mounds of North America that have so logically been attributed to the Mound Builders. In a letter, published in the *Astronomical Register*, 20-167, Mr. Birmingham calls attention to a formation that suggests the architectural upon the moon—"a group of three hills in a slightly acute-angled triangle, and connected by three lower embankments." There is a geometric object, or marking, shaped like an "X," in the crater Eratosthenes (*Sci. Amer. Sup.*, 59-24, 469); striking symbolic-looking thing or sign, or attempt by means of something obviously not topographic, to attract attention upon this earth, in the crater Plinius (*Eng. Mec.*, 35-34); reticulations, like those of a city's squares, in Plato (*Eng. Mec.*, 64-253); and there is a structural-looking composition of angular lines in Gassendi (*Eng. Mec.*, 101466). Upon the floor of Littrow are six or seven spots arranged in the form of the Greek letter Gamma (*Eng. Mec.*, 101-47). This arrangement may be of recent origin, having been discovered Jan. 31, 1915. The Greek letter makes difficulty only for those who do not want to think easily upon this subject. For a representation of something that looked like a curved wall upon the moon, see*L'Astronomie*, 1888-110. As to appearances like viaducts, see *L'Astronomie*, 1885-213. The lunar craters are not in all instances the simple cirques that they are commonly supposed to be. I have many different impressions of some of them: I remember one sketch that looked like an owl with a napkin tucked under his beak. However, it may be that the general style of architecture upon the moon is Byzantine, very likely, or not so likely, domed with glass, giving the dome-effect that has so often been commented upon.

So then the little nearby moon—and it is populated by Liliputians. However, our experience with agreeing ideas having been what it has been, we suspect that the lunarians are giants. Having reasonably determined that the moon is one hundred miles in diameter, we suppose it is considerably more or less.

A group of astronomers had been observing extraordinary lights in the lunar crater Plato. The lights had definite arrangement. They were so individualized that Birt and Elger, and the other selenographers, who had combined to study them, had charted and numbered them. They were fixed in position, but rose and fell in intensity.

It does seem to me that we have data of one school of communicationists after another coming into control of efforts upon the moon. At first our data related to single lights. They were extraordinary, and they seem to me to have been signals, but there seemed to be nothing of the organization that now does seem to be creeping into the fragmentary material that is the best that we can find. The grouped lights in Plato were so distinctive, so clear and even brilliant, that if such lights had ever shone before, it seems that they must have been seen by the Schroeters, Gruithuisens, Beers and Mädlers, who had studied and charted the features of the moon. For several of Gledhill's observations, from which I derive my impressions of these lights, see *Rept. B. A.*, 1871-80—"I can only liken them to the small discs of stars, seen in the transit-instrument"; "just like small stars in the transit instrument, upon a windy night!"

In August and September,. 1860, occurred a notable illumination of the spots in Group I. It was accompanied by a single light upon a distant spot.

February and March, 1870—illumination of another group.

April 17, 1870—another illumination in Plato, but back to the first group.

As to his observations of May 10-12, 1870, Birt gives his opinion that the lights of Plato were not effects of sunlight.

Upon the 13th of May, 1870, there was an "extraordinary display," according to Birt: 27 lights were seen by Pratt, and 28 by Elger, but only 4 by Gledhill, in Brighton. Atmospheric conditions may have made this difference, or the lights may have run up or down a scale from 4 to 28. As to independence of sunlight, Pratt says (*Rept. B. A.*, 1871-88) as to this display, that only the fixed, charted points so shone, and that other parts of the crater were not illuminated, as they would have been to an incidence common throughout. In Pratt's opinion, and, I think, in the opinion of the other observers, these lights were volcanic. It seems to me that this opinion arose from a feeling that there should be something of an opinion: the idea that the lights might have been signals was not expressed by any of these astronomers that I know of. I note that, though many observers were, at this time, concentrating upon this one crater, there are no records find-able by me of such disturbance of detail as might be supposed to accompany volcanic action. The clear little lights seem to me to have been anything but volcanic.

The play of these lights of Plato—their modulations and their combinations—like luminous music—or a composition of signals in a code that even in this late day may be deciphered. It was like orchestration—and that something like a baton gave direction to Light 22, upon Aug. 12, 1870, to shine a leading part —"remarkable increase of brightness." No. 22 subsided, and the leading part shone out in No. 14. It, too, subsided, and No. 16 brightened.

Perhaps there were definite messages in a Morse-like code. There is a chance for the electricity in somebody's imagination to start crackling. Up to April, 1871, the selenographers had recorded 1,600 observations upon the fluctuations of the lights of Plato, and had drawn 37 graphs of individual lights. All graphs and other records were deposited by W. R. Birt in the Library of the Royal Astronomical Society, where presumably they are to this day. A Champollion may some day decipher hieroglyphics that may have been flashed from one world to another.

21

Our data indicate that the planets are circulating adjacencies.

Almost do we now conceive of a difficulty of the future as being not how to reach the planets, but how to dodge them. Especially do we warn aviators away from that rhinoceros of the skies, Mercury. I have a note somewhere upon one of the wickedest-looking horns in existence, sticking out far from Mercury. I think it was Mr. Whitmell who made this observation. I'd like to hear Andrew Barclay's opinion upon that. I'd like to hear Capt. Noble's.

If sometimes does the planet Mars almost graze this earth, as is not told by the great telescopes, which are only millionaires' memorials, or, at least, which reveal but little more than did the little spy glasses used by Burnham and Williams and Beer and Mädler—but if periodically the planet Mars comes very close to this earth, and, if Mars, an island with perhaps no more surface-area than has England, but likely enough inhabited, like England—

June 19, 1875—opposition of Mars.

Flashes that were seen in the sky upon the 25th of June, 1875, by Charles Gape, of Scole, Norfolk (*Eng. Mec.*, 21-488). The Editor of *Symons' Met. Mag.* (see vol. 10-116) was interested, and sent Mr. Gape some questions, receiving answers that nothing had appeared in the local newspapers upon the subject, and that nothing could be learned of a display of fireworks, at the time. To Mr. Gape the appearances seemed to be meteoric.

The year 1877—climacteric opposition of Mars.

There were some discoveries.

We have at times wondered how astronomers spend their nights. Of course, according to many of his writings upon the subject, Richard Proctor had an excellent knowledge of whist. But in the year 1877, two astronomers looked up at the sky, and one of them discovered the moons of Mars, and the other called attention to lines on Mars—and, if for centuries, the moons of Mars could so remain unknown to all inhabitants of this earth except, as it were, Dean Swift—why, it is no wonder that we so respectfully heed some of the Dean's other intuitions, and think that there may be Liliputians, or Brobdingnagians, and other forms not conventionally supposed to be. As to our own fields of data, I have a striking number of notes upon signal-like appearances upon the moon, in the year 1877, but have notes upon only one occurrence that, in our interests, may relate to Mars. The occurrence is like that of July 31, 1813, and June 19, 1875.

Sept. 5, 1877—opposition of Mars.

Sept. 7, 1877—lights appeared in the sky of Bloomington, Indiana. They were supposed to be meteoric. They appeared and disappeared, at intervals of three or four seconds; darkness for several minutes; then a final flash of light. See *Sci. Amer.*, 37-193.

That all luminous objects that are seen in the sky when the planet Venus is nearest may not be Venus; may not be fire-balloons:

In the *Dundee Advertiser*, Dec. 22, 1882, it is said that, between 10 and 11 A.M., December 21, at Broughty Ferry, Scotland, a correspondent had seen an unknown luminous body near and a little above the sun. In the *Advertiser*, December 25, is published a letter from someone who says that this object had been seen at Dundee, also; that quite certainly it was the planet Venus and "no other." In *Knowledge*, 2-489, this story is told by a writer who says that undoubtedly the object was Venus. But, in *Knowledge*, 3-13, the astronomer J.

E. Gore writes that the object could not have been Venus, which upon this date was 1 h. 33 m., R. A., west of the sun. The observation is reviewed in *L'Astronomie*, 1883-109. Here it is said that the position of Mercury accorded better. Reasonably this object could not have been Mercury: several objections are comprehended in the statement that superior conjunction of Mercury had occurred upon December 16.

Upon Feb. 3, 1884, M. Staevert, of the Brussels Observatory, saw, upon the disc of Venus, an extremely brilliant point (*Ciel et Terre*, 5-127). Nine days later, Niesten saw just such a point of light as this, but at a distance from the planet. If no one had ever heard that such things cannot be, one might think that these two observations were upon something that had been seen leaving Venus and had then been seen farther along. Upon the 3rd of July, 1884, a luminous object was seen moving slowly in the sky of Norwood, N. Y. It had features that suggest the structural: a globe the size of the moon, surrounded by a ring; two dark lines crossing the nucleus (*Science Monthly*, 2-136). Upon the 26th of July, a luminous globe, size of the moon, was seen at Cologne; it seemed to be moving upward from this earth, then was stationary "some minutes," and then continued upward until it disappeared (*Nature*, 30-360). And in the *English Mechanic*, 40-130, it is not said that a luminous vessel that had sailed out from Venus, in February, visiting this earth, where it was seen in several places, was seen upon its return to the planet, but it is said that an observer in Rochester, N. Y., had, upon August 17, seen a brilliant point upon Venus.

22

Explosions over the towns of Barisal, Bengal, if they were aërial explosions, were continuing. As to some of these detonations that were heard in May, 1874, a writer in *Nature*, 53-197, says that they did seem to come from overhead. For a report upon the Barisal Guns, heard between April 28,

1888, and March 1, 1889, see *Proc. Asiatic Soc. of Bengal*, 1889-199.

Phenomena at Comrie were continuing. The latest date in Roper's *List of Earthquakes* is April 8, 1886, but this list goes on only a few years later. See *Knowledge*, n. s., 6-145—shock and a rumbling sound at Comrie, July 12, 1894—a repetition upon the corresponding date, the next year. In the *English Mechanic*, 74-155, David Packer says that, upon Sept. 17, 1901, ribbon-like flashes of lightning, which were not ordinary lightning, were seen in the sky (I think of Birmingham) one hour before a shock in Scotland. According to other accounts, this shock was in Comrie and surrounding regions (London *Times*,

Sept. 19, 1901).

Smithson. Miscell. Cols., 37-Appendix, p. 71:

According to L. Tennyson, Quartermaster's Clerk, at Fort Klamath, Oregon, at daylight, Jan. 8, 1867, the garrison was startled from sleep by what he supposed to be an earthquake and a sound like thunder. Then came darkness, and the sky was covered with black smoke or clouds. Then ashes, of a brownish color, fell—"as fast as I ever saw it snow." Half an hour later there was another shock, described as "frightful." No one was injured, but the sutler's store was thrown a distance of ninety feet, and the vibrations lasted several minutes. Mr. Tennyson thought that somewhere near Fort Klamath, a volcano had broken loose, because, in the direction of the Klamath Marsh, a dark column of smoke was seen. I can find record of no such volcanic eruption. In a list of quakes, in Oregon, from 1846, to 1916, published in the *Bull. Seis. Soc. Amer.*, September, 1919, not one is attributed to volcanic eruptions. Mr. W. D. Smith, compiler of the list, says, as to the occurrence at Fort Klamath—"If there was an eruption, where was it?" He asks whether possibly it could have been in Lassen Peak. But Lassen Peak is in California,, and the explosion upon Jan. 8, 1867, was so close to Fort Klamath that almost immediately ashes fell from the sky.

The following is of the type of phenomena that might be considered evidence of signaling from some unknown world nearby:

La Nature, 17-126—that, upon June 17, 1881, sounds like cannonading were heard at Gabes, Tunis, and that quaking of the earth was felt, at intervals of 32 seconds, lasting about 6 minutes.

July 30, 1883—a somewhat startling experience—steamship *Resolute* alone in the Arctic Ocean—six reports like gunfire—*Nature*, 53-295.

In *Nature*, 30-19, a correspondent writes that, upon the 3rd of January, 1869, a policeman in Harlton, Cambridgeshire, heard six or seven reports, as if of heavy guns far away. There is no findable record of an earthquake in England upon this date. In the London *Times*, Jan. 12, 15, 16, 1869, several correspondents write that upon the 9th of January a loud report had been heard and a shock felt at places near Colchester, Essex, about 30 miles from Harlton. One of the correspondents writes that he had heard the sound but had felt no shock. In the London *Standard*, January 12, the Rev. J. F. Bateman, of South Lopham, Norfolk, writes as to the occurrence upon the 9th—"An extraordinary vibration (described variously by my parishioners as being `like a gunpowder explosion,' 'a big thunder clap,' and 'a little earthquake') was noticed here this morning about 11.20." In the *Morning Post*, January 14, it is said that at places about twenty miles from Colchester it was thought that an explosion had occurred, upon the 9th, but, inasmuch as no explosion had been heard of, the disturbance was attributed to an earthquake. Night of January 13 —an explosion in the sky, at Brighton (*Rept. B. A.*, 1869-307). In the *Standard*, January 22, a correspondent writes from Swaffham, Norfolk, that, about 8 P.M., January 15, something of an unknown nature had frightened flocks of sheep, which had burst. from their bounds in various places. All these occurrences were in adjoining counties in southeastern England. Something was seen in the sky upon the 13th, and, according to the *Chudleigh Weekly Express*, Jan. 13, 1869, something was seen in the sky, night of the 10th, at Weston-super-Mare, near Bristol, in southwestern England. It was seen between 9 and 10 o'clock, and is said to have been an extraordinary meteor. Five hours later were felt three shocks said to have been earthquakes.

Upon the night of March 17, 1871, there was a series of events in France, and a series in England. A "meteor" was seen at Tours, at 8 P.M.—at 10:45, a "meteor" that left a luminous cloud over Saintes (Charante-Inferieure)— another at Paris, 11:15, leaving a mark in the sky, of fifteen minutes' duration —another at Tours, at 11:45 P.M. See *Les Mondes*, 24-190, and *Comptes Rendus*, 72-789. There were "earthquakes" this night affecting virtually all England north of the Mersey and the Trent, and also southern parts of Scotland. As has often been the case, the phenomena were thought to have been explosions and were then said to have been earthquakes when no terrestrial explosions could be heard of (*Symons' Met. Mag.*, 6-39). There were six shocks near Manchester, between 6 and 7 P.M., and others about 11 P.M.; and in Lancashire about 11 P.M., and continuing in places as far apart as Liverpool and Newcastle, until 11:30 o'clock. The shocks felt about u o'clock correspond, in time, with the luminous phenomena in the sky of France, but our way of expressing that these so-called earthquakes in England may have been concussions from repeating explosions in the sky, is to record that, according to correspondence in the London *Times*, there were, upon the 20th, aërial phenomena in the region of Lancashire that had been affected upon the 17th—"sounds that seemed to come from a number of guns at a distance" and "pale flashes of lightning in the sky."

Whether these series of phenomena be relatable to Mars or Martians or not, we note that in 1871 opposition of Mars was upon March 19; and, in 1869, upon February 13; and in 1867 two days after the explosions at Fort Klamath. In our records in this book, similar coincidences can be found up to the year 1879. I have other such records not here published, and others that will be here investigated.

There is a triangular region in England, three points of which appear so often in our data that the region should be specially known to us, and I know it myself as the London Triangle. It is pointed in the north by Worcester and Hereford, in the south by Reading, Berkshire, and in the east by Colchester, Essex. The line between Colchester and Reading runs through London.

Upon Feb. 18, 1884, at West Mersea, near Colchester, a loud report was heard (*Nature*, 53-4). Upon the 22nd of April, 1884, centering around Colchester, occurred the severest earthquake in England in the 19th century. For several columns of description, see the London *Times*, April 23. There is a long list of towns in which there was great damage: in 24 parishes near Colchester, 1,250 buildings were damaged. One of the places that suffered most was West Mersea (*Daily Chronicle*, April 28).

There was something in the sky. According to G. P. Yeats (*Observations upon the Earthquake of Dec. 17, 1896*, p. 6) there was a red appearance in the sky over Colchester, at the time of the shock of April 22, 1884.

The next day, according to a writer in *Knowledge*, 5-336, a stone fell from the sky, breaking glass in his greenhouse, in Essex. It was a quartz stone, and unlike anything usually known as meteoritic.

The indications, according to my reading of the data, and my impressions of such repeating occurrences as those at Fort Klamath, are that perhaps an explosion occurred in the sky, near Colchester, upon Feb. 18, 1884; that a great explosion did occur over Colchester, upon the 22nd of April, and that a great volume of débris spread over England, in a northwesterly direction, passing over Worcestershire and Shropshire, and continuing on toward Liverpool, nucleating moisture and falling in blackest of rain. From the Stonyhurst Observatory, near Liverpool, was reported, occurring at a 11 A.M., April 26, "the most extraordinary darkness remembered"; forty minutes later fell rain "as black as ink," and then black snow and black hail (*Nature*, 30-6). Black hail fell at Chaigley, several miles from Liverpool (*Stonyhurst Magazine*, 1-267). Five hours later, black substance fell at Crowle, near Worcester (*Nature*, 30-32). Upon the 28th, at Church Stretton and Much Wenlock, Shropshire, fell torrents of liquid like ink and water in equal proportions (*The Field*, May 3, 1884). In the *Jour. Roy. Met. Soc.*, 11-7, it is said that, upon the 28th, half a mile from Lilleshall, Shropshire, an unknown pink substance was brought down by a storm. Upon the 3rd of May, black substance fell again at Crowle (*Nature*, 30-32) .

In *Nature*, 30-216, a correspondent writes that, upon June 22, 1884, at Fletching, Sussex, southwest of Colchester, there was intense darkness, and that rain then brought down flakes of soot in such abundance that it seemed to be "snowing black." This was several months after the shock at Colchester, but my datum for thinking that another explosion, or disturbance of some kind, had occurred in the same local sky, is that, as reported by the inmates of one house, a slight shock was felt, upon the 24th of June, at Colchester, showing that the phenomena were continuing. See Roper's *List of Earthquakes*.

Was not the loud report heard upon February 18 probably an explosion in the sky, inasmuch as the sound was great and the quake little? Were not succeeding phenomena sounds and concussions and the fall of débris from explosions in the sky, acceptably upon April 22, and perhaps continuing until the 24th of June? Then what are the circumstances by which one small part of this earth's surface could continue in relation with something somewhere else in space?

Comrie, Irkutsk, and Birmingham.

23

Upon the night of the 13th of July, 1875, at midnight, two officers of H.M.S. *Coronation*, in the Gulf of Siam, saw a luminous projection from the moon's upper limb (*Nature*, 12-495) . Upon the 14th it was gone, but a smaller projection was seen from another part of the moon's limb. This was in the period of the opposition of Mars.

Upon the night of Feb. 20, 1877, M. Trouvelot, of the Observatory of Meudon, saw, in the lunar crater Eudoxus, which, like almost all other centers of seeming signaling, is in the northwestern quadrant of the moon, a fine line of light (*L'Astronomie*, 1885-212). It was like a luminous cable drawn across the crater.

March 21, 1877—a brilliant illumination, and not by the light of the sun, according to C. Barrett, in the lunar crater Proclus (*Eng. Mec.*, 25-89).

May 15 and 29, 1877-the bright spot west of Picard (*Eng. Mec.*, 25-335).

The changes upon Linné were first seen by Schmidt, in 1866, near the time of opposition of Mars. In May, 1877, Dr. Klein announced that a new object had appeared upon the moon. It was close to the center of the visible disc of the moon, and was in a region that had been most carefully studied by the selenographers. In the *Observatory*, 2-238, is Neison's report from his own memoranda. In the years 1874 and 1875, he had studied this part of the moon, but had not seen this newly reported object in the crater Hyginus, or the object, Hyginus N, according to the selenographers' terminology. In the *Astronomical Register*, 17-204, Neison lists, with details, 20 minute examinations of this region, from July, 1870, to August, 1875, in which this conspicuous object was not recorded.

June 14, 1877—a light on the dark part of the moon, resembling a reflection from a moving mirror; reported by Prof. Henry Harrison (*Sidereal Messenger*, 3-150). June 15—the bright spot west of Picard, according to Birt (*Jour. B. A. A.*, 19-376). Upon the 16th, Prof. Harrison thought that again he saw the moving light of the 14th, but shining faintly. In the *English Mechanic*, 25-432, Frank Dennett writes, as to an observation of June 17, 1877—"I fancied I could detect a minute point of light shining out of the darkness that filled Bessel."

These are data of extraordinary activity upon the moon preceding the climacteric opposition of Mars, early in September, 1877. Now we have an account of an occurrence during an eclipse of the moon:

On the night of the eclipse (Aug. 27, 1877) a ball of fire, of the apparent size of the moon, was seen, at ten minutes to eleven, dropping apparently from cloud to cloud, and the light flashing across the road (*Astro. Reg.*, 1878-75).

Astro. Reg., 17-251:

Nov. 13, 1877—Hyginus N standing out with such prominence as to be seen at the first glance;

Nov. 14, 1877—not a trace of Hyginus N, though seeing was excellent:

Oct. 3, 1878—the most conspicuous of all appearances of Hyginus N;

Oct. 4, 1878—not a trace of Hyginus N.

Upon the night of Nov. 1, 1879, again in the period of opposition of Mars (opposition November 12) again the bright spot west of Picard (*Jour B. A. A.*, 19-376). But I have several records of observations upon this appearance not in times of opposition of Mars. Whether there be any relation with anything else or not, at five o'clock, morning of Nov. 1, 1879, a "vivid flash" was seen and a shock was felt at West Cumberland (*Nature*, 21-19).

In the autumn of the year 1883, began extraordinary atmospheric effects in the sky of this earth. For Prof. John Haywood's description of similar appearances upon the moon, Nov. 4, 1883, and March 29, 1884, see the *Sidereal Messenger*, 3-121. They were misty light-effects upon the dark part of the moon, not like "earth-shine." Our expression is that so close is the moon to this earth that it, too, may be affected by phenomena in the atmosphere of this earth.

Something like another luminous cable, or like a shining wall, that was seen in Aristarchus, by Trouvelot, Jan. 23, 1880 (*L'Astro.*, 1885-215); a speck of light in Marius, Jan. 13, 1881, by A. S. Williams (*Eng. Mec.*, 32-494); unexplained light in Eudoxus, by Trouvelot, May 4, 1881 (*L'Astro.*, 1885-213); an illumination in Kepler, by Morales, Feb. 5, 1884 (*L'Astro.*, 9-149).

In Knowledge, 7-224, William Gray writes that, upon Feb. 19, 1885, he saw, in Hercules, a dull, deep, reddish appearance. In*L'Astronomie*, 1885-227, Lorenzo Kropp, an astronomer of Paysandu, Uruguay, writes that, upon Feb. 21, 1885, he had seen, in Cassini, a formation not far from Hercules, both of them in the northwestern quadrant of the moon, a reddish smoke or mist. He had heard that several other persons had seen, not a misty appearance, but a star-like light here, and upon the 22nd he had seen a definite light, himself, shining like the planet Saturn.

May 11, 1885—two lights upon the moon (*L'Astro.*, 9-73).

May 11, 1886—two lights upon the moon (*L'Astro.*, 6-312).

24

THAT through lenses rimmed with horizons, inhabitants of this earth have seen revelations of other worlds—that atmospheric strata of different densities are lenses—but that the faults of the wide glasses in the observatories are so intensified in atmospheric revelations that all our data are distortions. Our acceptance is that every mirage has a primary; that in human mind all poetry is based upon observation, and that imagery in the sky is similarly uncreative. If a mirage cannot be traced to the known upon this earth, one supposes that it is either a derivation from the unknown upon this earth, or from the unknown somewhere else. We shall have data of a series of mirages in Sweden, or upon the shores of the Baltic, from October, 1881, to December, 1888. I take most of the data from *Nature, Knowledge, Cosmos,* and *L'Astronomie,* published in this period. I have no data of such appearances in this region either before or after this period: the suggestion in my own mind is that they were not mirages from terrestrial primaries, or they would not be so confined to one period, but were shadows or mirages from something that was in temporary suspension over the Baltic and Sweden, all details distorted and reported in terms of familiar terrestrial appearances.

Oct. 10, 1881—that at Rugenwalde, Pomerania, the mirage of a village had been seen: snow-covered roofs from which hung icicles; human forms distinctly visible. It was believed that the mirage was a representation of the town of Nexo, on the island of Bornholm. Rugenwalde is on the Baltic, and Nexo is about 100 miles northwest, in the Baltic.

The first definite account of the mirages of Sweden, findable by me, is published in *Nature,* June 29, 1882, where it is said that preceding instances had attracted attention—that, in May, 1882, over Lake Orsa, Sweden, representations of steamships had been seen, and then "islands covered with vegetation." Night of May 19, 1883—beams of light at Lake Ludyika, Sweden —they looked like a representation of a lake in moonshine, with shores covered with trees, showing faint outlines of farms (*Monthly Weather Review*, May, 1883). May 28, 1883—at Finsbo, Sweden—changing scenes, at short intervals: mountains, lakes, and farms. Oct. 16, 1884—Lindsberg—a large town, with four-storied houses, a castle and a lake. May 22, 1885—Gothland —a town surrounded by high mountains, a large vessel in front of the town. June 15, 1885—near Oxelosund—two wooded islands, a construction upon one of them, and two warships. It is said that at the time two Swedish warships were at sea, but were at considerable distance north of Oxelosund. Sept. 12, 1885—Valla—a representation that is said to have been a "remarkable mirage" but that is described as if the appearances were cloud-forms—several monitors, one changing into a spouting whale, and the other into a crocodile— then forests—dancers—a wooded island with buildings and a park. Sept. 29, 1885—again at Valla—between 8 and 9 o'clock, P.M.; a lurid glare upon the northwestern horizon; a cloud bank—animals, groups of dancers, a forest, and then a park with paths. July 15, 1888—Hudikwall—a tempestuous sea, and a vessel upon it; a small boat leaving the vessel. Upon Oct. 8, 1888, at Merexull, on the Baltic, but in Russia, was seen a mirage of a city that lasted an hour. It is said that some buildings were recognized, and that the representation was identified with St. Petersburg, which is about 200 miles from the Baltic.

That a large, substantial mass, presumably of land, can be in at least temporary suspension over a point upon this earth's surface, and not fall, and be, in ordinary circumstances, invisible—

In *L'Astronomie*, 1887-426, MM. Codde and Payan, both of them astronomers, well-known for their conventional observations and writings, publish accounts of an unknown body that appeared upon the sun's limb, for twenty or thirty seconds, after the eclipse of Aug. 19, 1887. They saw a round body, apparent diameter about one tenth of the apparent diameter of the sun, according to the sketch that is published. In *L'Astronomie*, these two observers write separately, and, in the city of Marseilles, their observations were made at a distance apart. But the unknown body was seen by both upon the same part of the sun's limb. So it is supposed that it could not have been a balloon, nor a circular cloud, nor anything else very near this earth. But many astronomers in other parts of Europe were watching this eclipse, and it seems acceptable that others, besides two in Marseilles, continued to look, immediately after the eclipse; but from nowhere else came a report upon this object, so that all indications are that it was far from the sun and near Marseilles, but farther than clouds or balloons in this local sky. I can draw no diagram that can satisfy all these circumstances, except by supposing the sun to be only a few thousand miles away.

If little black stones fall four times, in eleven years, to one part of this earth's surface, and fall nowhere else, we are, in conceiving of a fixed origin somewhere above a stationary earth, at least conceiving in terms of data, and, whether we are fanatics or not, we are not of the type of other upholders of stationariness of this earth, who care more for Moses than they do for data. I'd not like to have it thought that we are not great admirers of Moses, sometimes.

The rock that hung in the sky of Servia—

Upon Oct. 13, 1872, a stone fell from the sky, to this earth, near the town of Soko-Banja, Servia. If it were not a peculiar stone, there is no force to this datum. It is said that it was unknown stone. A name was invented for it. The stone was called *banjite*, after the town near which it fell.

Seventeen years later (Dec. 1, 1889) another rock of *banjite* fell in Servia, near Jelica.

For Meunier's account of these stones, see *L'Astronomie*, 1890272, and *Comptes Rendus*, 92-331. Also, see *La Nature*, 1881-1-192. According to Meunier these stones did fall from the sky; indigenous to this earth there are no such stones; nowhere else have such stones fallen from the sky; they are identical in material; they fell seventeen years apart.

At times when we think favorably of this work of ours, we see in it a pointing-out of an evil of modern specialization. A seismologist studies earthquakes, and an astronomer studies meteors;. neither studies both earthquakes and meteors, and consequently each, ignorant of the data collected by the other, sees no relation between the two phenomena. The treatment of the event in Servia, Dec. 1, 1889, is an instance of conventional scientific attempts to understand something by separately, or specially, focusing upon different aspects, and not combining into an inclusive concept. Meunier writes only upon the stones that fell from the sky, and does not mention an earthquake at the time. Milne, in his *Catalogue of Destructive Earthquakes*, lists the occurrence as an earthquake, and does not mention stones that fell from the sky. All combinations greatly affect the character of components: in our combination of the two aspects, we see that the phenomenon was not an earthquake, as earthquakes are commonly understood, though it may have been meteoric; but was not meteoric, in ordinary terms of meteors, because of the unlikelihood that meteors, identical in material, should, seventeen years apart, fall upon the same part of this earth's surface, and nowhere else.

This occurrence was of course an explosion in the sky, and its vibrations were communicated to the earth below, with all the effects of any other kind of earthquakes. Back in our earliest confusion of the data of a century's first quarter, we had awareness of this combination and its conventional misinterpretation: that many concussions that have been communicated from explosions in the sky have been catalogued in lists of subterranean earthquakes. We are farther along now, in our data of the 19th century, and now we come across awareness, in other minds, of this distinguishment. At 8:20 A.M., Nov. 20, 1887, was heard and felt something that was reported from many places in the region that is known to us as the London Triangle, as an earthquake, though in some towns it was thought that a great explosion, perhaps in London, had occurred. It was reported from Reading, and from four towns near Reading, and Reading is said to be one of the places where the concussion was greatest. There were several accounts of slight alarm among sheep, which are sensitive to meteors and earthquakes. But, in *Symons' Met. Mag.*, Mr. H. G. Fordham wrote that the occurrence was not an earthquake; that a meteor had exploded. He had very little to base this opinion upon: out of scores of descriptions, he had record of only two assertions that something had been seen in the sky. Nevertheless, because the sound was so much greater than the concussion, Mr. Fordham came to his conclusion.

In *Symons' Met. Mag.*, 23-154, Dr. R. H. Wake writes that, upon the evening of Nov. 3, 1888, in a region about four miles wide and ten or fifteen miles long, in the Thames Valley (near Reading) flocks of sheep had rushed from their folds in a common alarm. About a year later, in the Chiltern Hills, which extend in a northeasterly direction from the Thames Valley, near Reading, there was another such occurrence. In the London *Standard*, Nov. 7, 1889, the Rev. J. Ross Barker, of Chesham, a town about 25 miles northeast of Reading, writes that, upon Oct. 25, 1889, many flocks of sheep, in a region of 30 square miles, had, by common impulse, broken from their folds. Mr. Barker asks whether anyone knew of a meteor or of an earthquake at the time. In vol. 24, *Symons' Met. Mag.*, Mr. Symons accepts that all three of these occurrences were effects of meteoric explosions in the sky. The phenomena are insignificant relatively to some that we have considered: the significance is in this definite recognition in orthodoxy, itself, that some supposed earthquakes, or effects of supposed earthquakes, are reactions to explosions in the sky.

25

Exploding monasteries that shoot out clouds of monks into cyclonic formations with stormy nuns similarly dispossessed —or collapsing monasteries— sometimes slowly crumbling confines of the cloistered—by which we typify all things: that all developments pass through a process of walling-away within shells that will break. Once upon a time there was a shell around the United States. The shell broke. Some other things were smashed.

The doctrines of great distances among heavenly bodies, and of a moving earth are the strongest elements of Exclusionism: the mere idea of separations by millions of miles discourages thoughts of communication with other worlds; and only to think that this earth shoots through space at a velocity of 19 miles a second puts an end to speculation upon how to leave it and how to return. But, if these two conventions be features of a walling-away like that of a chick within its shell, or that of the United States within its boundaries, and if some day all such confinements of the embryonic break, our own prophecy, in the vague terms of all successful prophecies, is that a matured view of astronomic phenomena will be from a litter of broken demonstrations.

Our expression now is upon the function of Isolation in Development. Specially it is not ours, because I think we learned it from the biologists, but we are applying it generally. If the general expression be accepted, we conceive that functionally have the astronomers taught that planets are millions of miles away, and that this earth moves at such terrific velocity that it is encysted with speed. Whether isolations function or not, that exclusions that break down are typical of all developments is signified by data upon all growing things, beginning with the aristocratic seeds, which, however, liberalize to intercourse with mean materials or die. All animal-organisms are at first walled away. In human circumstances conditions are the same. The development of every science has been a series of temporary exclusions, and the story of every industry tells of inventions that were resisted, but that were finally admitted. At the beginning of the nineteenth century, Hegel published his demonstration that there could be only seven planets: too late to recall the work, he learned that Ceres had been discovered. It is our expression that the mental state of Hegel partook of a general spirit of his time, and that it was necessary, or that it functioned, because early astronomers could scarcely have systematized their doctrine had they been bewildered by seven or eight hundred planetary bodies; and that, besides the functions of the astronomers, according to our expressions, there was also their usefulness in breaking down the walls of the older, and outlived, orthodoxy. We conceive that it is well that a great deal of experience should be withheld from children, and that, any way, in their early years, they are sexually isolated, for instance, and our idea is that our data have been held back by no outspoken conspiracy, but by an inhibition similar to that by which a great deal of biology, for instance, is not taught to children. But, if we think of something of this kind, equally acceptable is it that even in the face of orthodox principles, these data have been preserved in orthodox publications, and that, in the face of supposed principles of Darwinism, as applied generally they have survived, though not in harmony with their environment.

Tons of paper have been consumed by calculations upon the remoteness of stars and planets. But I can find nothing that has been calculated, or said, that is sounder than Mr. Shaw's determination that the moon is 37 miles away. It is that the Vogels and the Struves and the Newcombs have been functionally hypnotized and have usefully spread the embryonic delusion that there is a vast, untraversible expanse of space around this earth, or that they have had some basis that it has been my misfortune to be unable to find, or that there is no pleasant and unaccusatory way of explaining them.

April 10, 1874—a luminous object that exploded in the sky of Kuttenberg, Bohemia. It is said that the glare was like sunlight, and that the "terrifying flash" was followed by a detonation that rumbled about a minute. April 9, 1876—an explosion that is said to have been violent, over the town of Rosenau, Hungary. See *Rept. B. A.*, 1877-147.

These two objects which appeared in virtually the same local sky of this earth —points of explosion 250 miles apart—came from virtually the same point in the sky: constellation of Cassiopeia; different by two degrees in right ascension, and with no difference in declination. About the same time in the evening: one at 8:09 P.M., and the other at 8:20 P.M. Same night in the year, according to extra-terrestrial calendars: the year 1876 was a leap year.

If they had been ordinary meteors, by coincidence two ordinary meteors of the same stream might, exactly two years apart, come from almost the same point in the heavens and strike almost the same point over this earth. But they were two of the most extraordinary occurrences in the records of explosions in the sky. Coincidences multiply, or these objects did come from the not far-distant constellation Cassiopeia, and their striking so closely together indicates that this earth is stationary; and something of the purposeful may be thought of. Serially related to these events, or representing some more coincidence, there had been, upon June 9, 1866, a tremendous explosion in the sky of Knysahinya, Hungary, and about a thousand stones had fallen from the sky (*Rept. B. A.*, 1867-430). Rosenau and Knysahinya are about 75 miles apart. Of course one can very much extend our own circumscribed little notions, and think of the firing of projectiles from beyond the stars, just as one can think of our unknown lands as being not in the immediate sky of Servia or Birmingham or Comrie, but as being beyond the nearby stars, reducing everything more than we have reduced—but the firing of stones to this earth seems crude to me. Of course, objects, or fragments of objects made of steel, like the manufactured steel of this earth, have fallen to this earth, and are now in collections of "meteorites." There is a story in a book that is not very accessible to us, because it can't be found along withC. R., or *Eng. Mec.*, or *L'Astro.*, of tablets of stone that were once upon a time fired to this earth. It may be that inhabitants of this earth have been receiving instructions ever since, engravings arriving very badly damaged, however.

I have data upon repeating appearances, said to have been "auroral," in a local sky. If they were auroral, repetitions at regular intervals and so localized are challengers to the most resolute of explainers. If they were of extra-mundane origin, they indicate that this earth is stationary. The regularity is suggestive of signaling. For instance—a light in the sky of Lyons, N. Y., Dec. 9, 1891, Jan. 5, Feb. 2, Feb. 29, March 27, April 23, 1892. In the *Scientific American*, May 7, 1892, Dr. M. A. Veeder writes that, from Dec. 9, 1891, to April 23, 1892, there had been a bright light that he calls "auroral" in the sky of Lyons, every 27th night. He associates the lights with the sun's synodic period, and says that upon each of the days preceding a nocturnal display, there had been a disturbance in the sun. How a disturbance in the sun could, at night, sun somewhere near the antipodes of Lyons, N. Y., so localize its effects, one can't clear up. In *Nature*, 46-29, Dr. Veeder associates the phenomena with the synodic period of the sun, but he says that this period is of 27 days, 6 hours, and 47 minutes, noting that this period is inconsistent with the phenomena at Lyons, making more than a day's difference in the time of his records. This precise determination is more of the "exact science" that is driving some of us away from refinements into hoping for caves. Different parts of the sun move at different rates: I have read of sun spots that moved diagonally across the sun.

In *Nature*, 15-451, a correspondent writes that, at 8:55 P.M., he saw a large red star in Serpens, where he had never seen such an appearance before— Gunnersbury, March 17, 1877. Ten minutes later, the object increased and decreased several times, flashing like the revolving light of a lighthouse, then disappearing. This correspondent writes that, about Jo P.M., he saw a great meteor. He suggests no relation between the two appearances, but there may have been relation, and there may be indication of something that was stationary at least one hour over Gunnersbury, because the object said to have been a "meteor" was first seen at Gunnersbury. In the *Observatory*, 1-20, Capt. Tupman writes that, at 9:57 o'clock, a great meteor was seen first at Frome, Tetbury, and Gunnersbury. The red object might not have been in the local sky of Gunnersbury; might have been in the constellation Serpens, unseen in all the rest of the world.

There is a great field of records of "meteors" that, with no parallax, or with little parallax, or with little parallax that may be accounted for by supposing that observations were not quite simultaneous, have been seen to come as if from a star or from a planet, and that may have come from such points, indicating that they are not far away. For instance, *Rept. B. A.*, 1879-77—the great meteor of Sept. 5, 1868. It was seen, at Zurich, Switzerland, to come from a point near Jupiter; at Tremont, France, origin was so close to Jupiter that this object and the planet were seen in the same telescopic field; at Bergamo, Italy, it was seen five or six degrees from Jupiter. Zurich is about 140 miles from Bergamo, and Tremont is farther from Zurich and Bergamo than that.

So there are data that indicate that objects have come to this earth from planets or from stars, enforcing our idea that the remotest planet is not so far from this earth as the moon is said, conventionally, to be; and that the stars, all equidistant from this earth might be reached by traveling from this earth. One notices that I always conclude that, if phenomena repeatedly occur in one local sky of this earth, their origin is traceable to a fixed place over a stationary earth. The fixed place over this earth is indicated, but that fixed place—island of space, foreign coast, whatever it may be—may be conceived of as accompanying this earth in its rotations and revolutions around the sun. Accepting that nothing much is known of gravitation; that gravitational astronomy is a myth; that attraction may extend but a few miles around this earth, if I can think of something hanging unsupported in space, I always think of an island, say, over Birmingham, or Irkutsk, or Comrie, as soon flying off by the centrifugal force of a rotating earth, or as being soon left behind in a rush around the sun. Nevertheless there is good room for discussion here. But when it comes to other orders of data, I find one convergence toward the explanation that this earth is stationary. But the subject is supposed to be sacred. One must not think that this earth is stationary. One must not investigate. To think upon this subject, except as one is told to think, is, or seems to be considered, impious.

But how can one account for an earth that moves?

By thinking that something started it and that nothing ever stopped it.

Earth that doesn't move?

That nothing ever started it.

Some more sacrilege.

26

If a grasshopper could hop on a cannon ball, passing overhead, I could conceive, perhaps, how something, from outer space, could flit to a moving earth, explore a while, and then hop off.

But suppose we have to accept that there have been instances of just such enterprise and agility, relatively to the planet Venus. Irrespective of our notion that it may be that sometimes a vessel sails to this earth from Venus and returns, there are striking data indicating that, whether conceivable or not, luminous objects have appeared from somewhere, or presumably from outer space, and have been seen temporarily suspended over the planet Venus. This is in accord with our indications that there are regions in the sky suspended over and near this earth. It looks bad for our inference that this earth is stationary, but it is the supposed rotary motion of this earth more than the supposed orbital motion that seems to us would dislodge such neighboring bodies; and all astronomers, except those who say that Venus rotates in about 24 hours, say that Venus rotates in about 224 days, a velocity that would generate little centrifugal force.

I have a note upon a determined luminosity that was bent upon Saturn, as its objective. In the *English Mechanic*, 63-496, a correspondent writes that, upon July 13, 1896, he saw, through his telescope, from 10 until after 11:15 P.M., after which the planet was too near the horizon for good seeing, a luminous object moving near Saturn. He saw it pass several small stars. "It was certainly going toward Saturn at a good rate." There may be swifts of the sky that can board planets. If they can swoop on and off an earth moving at a rate of 19 miles a second, disregarding rotation, because entrance at a pole may be thought of, why, then, for all I know smaller things do ride on cannon balls. Of course if our data that indicate that the supposed solar system, or the geo-system, is to an enormous degree smaller than is conventionally taught be accepted, the orbital velocity of Venus is far cut down.

About the last of August, 1873—Brussels; eight o'clock in the evening—rising above the horizon, into a clear sky, was seen a star-like object. It mounted higher and higher, until, about ten minutes later, it disappeared (*La Nature*, 1873-239). It seems that this conspicuous object did appear in a local sky, and was therefore not far from this earth. If it were not a fire-balloon, one supposes that it did come from outer space, and then returned.

Perhaps a similar thing that visited the moon, and was then seen sailing away —in the *Astronomical Register*, 23-205, Prof. Schafarik, of Prague, writes that upon April 24, 1874, he saw "an object of so peculiar a nature that I do not know what to make of it." He saw a dazzling white object slowly traversing the disc of the moon. He had not seen it approaching the moon. He watched it after it left the moon. Sept. 27, 1881—South Africa—an object that was seen near the moon, by Col. Markwick—like a comet but moving rapidly (*Jour. Liverpool Astro. Soc.*, 7-117).

Our chief interest is in objects, like ships, that have "boarded" this moving earth with the agility of a Columbus who could dodge a San Salvador and throw out an anchor to an American coast screeching past him at a rate of 19 miles a second, or in objects that have come as close as atmospheric conditions, or unknown conditions, would permit to the bottom of a kind of stationary sea. We now graduate Capt. Noble to the extra-geographic fold. In *Knowledge*, 4-173, Capt. Noble writes that, at 10:35 o'clock, night of Aug. 28, 1883, he saw in the sky something "like a new and most glorious comet." First he saw something like the tail of a comet, or it was like a searchlight, according to Capt. Noble's sketch of it in *Knowledge*. Then Capt. Noble saw the nucleus from which this light came. It was a brilliant object. Upon page 207, W. K. Bradgate writes that, at 12:40 A.M., August 29, at Liverpool, he saw an object like the planet Jupiter, a ray of light emanating from it. Upon the nights of September 11 and 13, Prof. Swift saw, at Rochester, N. Y., an unknown object like a comet, perhaps in the local sky of Rochester, inasmuch as it was reported from nowhere else (*Observatory*, 6-345). In *Knowledge*, 4-219, Mrs. Harbin writes that, upon the night of September 21, at Yeovil, she saw the same brilliant searchlight-like light that had been seen by Capt. Noble, but that it had disappeared before she could turn her telescope upon it. And several months later (November, 1883) a similar object was seen obviously not far away, but in the local sky of Porto Rico and then of Ohio (*Amer. Met. Jour.*, 1-110, and *Sci. Amer.*, 50-40, 97). It may be better not to say at this time that we have data for thinking that a vessel carrying something like a searchlight, visited this earth, and explored for several months over regions as far apart as England and Porto Rico. Just at present it is enough to record that something that was presumably not a fire-balloon appeared in the sky of England, close to this earth, if seen nowhere else, and in two hours traversed the distance of about 200 miles between Sussex and Liverpool.

Aug. 22, 1885—Saigon, Cochin-China—according to Lieut. Réveillère, of the vessel *Guiberteau*—object like a magnificent red star, but larger than the planet Venus—it moved no faster than a cloud in a moderate wind; observed 7 or 8 minutes, then disappearing behind clouds (*C. R.*, 101-680).

In this book it is my frustrated desire to subordinate the theme of this earth's stationariness. My subject is New Lands—things, objects, beings that are, or may be, the data of coming expansions

But the stationariness of this earth cannot be subordinated. It is crucial.

Again—there is no use discussing possible explorations beyond this earth, if this earth moves at a rate of 19 miles a second, or 19 miles a minute.

As to voyagers who may come to or near this earth from other planets—how could they leave and return to swiftly moving planets? According to our principles of Extra-geography, the planets move part of the time with the revolving stars, the remotest planets remaining in, under, or near one constellation years at a time. Anything that could reach, and then travel from, a swiftly revolving constellation in the ecliptic could arrive at a stellar polar region, .where, relatively to a central, stationary body, there is no motion.

27

It may be that we now add to our sins the horse that swam in the sky. For all I know, we contribute to a wider biology. In the *New York Times*, July 8, 1878, is published a dispatch from Parkersburg, West Virginia: that, about July 1, 1878, three or four farmers had seen, in a cloudless sky, apparently half a mile high, "an opaque substance." It looked like a white horse, "swimming in the clear atmosphere." It is said to have been a mirage of a horse in some distant field. If so, it is interesting not only because it was opaque, but because of a selection or preference: the field itself was not miraged.

Black bodies and the dark rabbles of the sky—and that rioting thing, from floating anarchies, have often spotted the sun. Then, by all that is compensatory, in the balances of existence, there are disciplined forces in space. In the *Scientific American*, 44-291, it is said that, according to newspapers of Delaware, Maryland, and Virginia, figures had been seen in the sky in the latter part of September, and the first week in October, 1881, reports that "exhibited a mediæval condition of intelligence scarcely less than marvelous." The writer suggests that, though probably something had been seen in the sky, it was only an aurora. Our own intelligence and that of astronomers and meteorologists and everybody else with whom we have had experience had better not be discussed, but the accusation of mediævalism is something that we're sensitive about, and we hasten to the *Monthly Weather Review*, and if that doesn't give us a modern touch, I mistake the sound of it. *Monthly Weather Review*, September and October, 1881—an auroral display in Maryland and New York, upon the 23rd of September; all other auroras in September far north of the three states in which it was said phenomena were seen. October—no auroras until the 18th; that one in the north. There was a mirage upon September 23, but at Indianola; two instances in October, but late in the month, and in northern states.

It is said, in the *Scientific American*, that, according to the *Warrentown* (Va.) *Solid South*, a number of persons had seen white-robed figures in the sky, at night. The story in the *Richmond Dispatch* is that many persons had seen, or had thought they had seen, an alarming sight in the sky, at night: a vast number of armed, uniformed soldiers drilling. Then a dispatch from Wilmington, Delaware—platoons of angels marching and countermarching in the sky, their white robes and helmets gleaming. Similar accounts came from Laurel and Talbot. Several persons said that they had seen, in the sky, the figure of President Garfield, who had died not long before. Our general acceptance is that all reports upon such phenomena are colored in terms of appearances and subjects uppermost in minds.

L'Astronomie, 1888-392:

That, about the first of August, 1888, near Warasdin, Hungary, several divisions of infantry, led by a chief, who waved a flaming sword, had been seen in the sky, three consecutive days, marching several hours a day. The writer in *L'Astronomie*says that in vain does one try to explain that this appearance was a mirage of terrestrial soldiers marching at a distance from Warasdin, because widespread publicity and investigation had disclosed no such soldiers. Even if there had been terrestrial soldiers near Warasdin repeating mirages localized would call for explanation.

But that there may be space-armies, from which reflections or shadows or Brocken specters are sometimes cast—a procession that crossed the sun: forms that moved, or that marched, sometimes four abreast; observation by M. Bruguière, at Marseilles, April 15 and 16, 1883 (*L'Astro.,* 5-70). An army that was watched, forty minutes, by M. Jacquot, Aug. 30, 1886 (*L'Astro.,* 1886-71) —things or beings that seemed to march and to counter-march: all that moved in the same direction, moved in parallel lines. In *L'Année Scientifique,* 29-8, there is an account of observations by M. Trouvelot, Aug. 29, 1871. He saw objects, some round, some triangular, and some of complex forms. Then occurred something that at least suggests that these things were not moving in the wind, nor sustained in space by the orbital forces of meteors; that each was depending upon its own powers of flight, and that an accident occurred to one of them. All of them, though most of the time moving with great rapidity, occasionally stopped, but then one of them fell toward the earth, and the indications are that it was a heavy body, and had not been sustained by the wind, which would scarcely suddenly desert one of its flotsam and continue to sustain all the others. The thing fell, oscillating from side to side like a disc falling through water.

New York Sun, March 16, 1890—that, at 4 o'clock, in the afternoon of March 12th, in the sky of Ashland, Ohio, was seen a representation of a large, unknown city. By some persons it was supposed to be a mirage of the town of Mansfield, thirty miles away; other observers thought that they recognized Sandusky, sixty miles away. "The more superstitious declared that it was a vision of the New Jerusalem."

May have been a revelation of heaven, and for all I know heaven may resemble Sandusky, and those of us who have no desire to go to Sandusky may ponder that point, but our own expression is that things have been pictured in the sky, and have not been traced to terrestrial origins, but have been interpreted always in local terms. Probably a living thing in the sky— seen by farmers—a horse. Other things, or far-refracted images, or shadows— and they were supposed to be vast lions or soldiers or angels, all according to preconceived ideas. Representations that have been seen in India—Hindoo costumes described upon them. Suppose that, in the afternoon of Jan. 17, 1892, there was a battle in the sky of Montana—we know just about in what terms the description would be published. *Brooklyn Eagle*, Jan. 18, 1892—a mirage in the sky of Lewiston, Montana—Indians and hunters alternately charging and retreating. The Indians were in superior numbers and captured the hunters. Then details—hunters tied to stakes; the piling of faggots; etc. "So far as could be ascertained last night, the Indians on the reservations are peaceable." I think that we're peaceable enough, but, unless the astronomers can put us on reservations, where we'll work out expressions in beads and wampum instead of data, we'll have to carry on a conflict with the vacant minds to which appear mirages of their own emptiness in the sometimes swarming skies.

Altogether there are many data indicating that vessels and living things of space do come close to this earth, but there is absence of data of beings that have ever landed upon this earth, unless someone will take up the idea that Kaspar Hauser, for instance, came to this earth from some other physical world. Whether spacarians have ever dredged down here or not, or "sniped" down here, pouncing, assailing, either wantonly, or in the interests of their sciences, there are data of seeming seizures and attacks from somewhere, and I have strong objections against lugging in the fourth dimension, because then I am no better off, wondering what the fifth and sixth are like.

In *La Nature*, 1888-2-66, M. Adrian Arcelin writes that, while excavating near de Solutré, in August, 1878, upon a day, described as *superbe*, sky clear to a degree said to have been *parfaitement*, several dozen sheets of wrapping paper upon the ground suddenly rose. Nearby were a dozen men, and not one of them had felt a trace of wind. A strong force had seized upon these conspicuous objects, touching nothing else. According to M. Arcelin, the dust on the ground under and around was not disturbed. The sheets of paper continued upward, and disappeared in the sky.

A powerful force that swooped upon a fishing vessel, raising it so far that when it fell back it sank—see London *Times*, Sept. 24, 1875. A quarter of a mile away were other vessels, from which set out rescuers to the sailors who had been thrown into the sea. There was no wind: the rescuers could not use sails, but had to row their boats.

Upon Oct. 2, 1875, a man was trundling a cart from Schaffhausen, near Beringen, Germany. His right arm was perforated from front to back, as if by a musket ball (*Pop. Sci.*, 15-566). This man had two companions. He had heard a whirring sound, but his companions had heard nothing. At one side of the road there were laborers in a field, but they were not within gunshot distance. Whatever the missile may have been, it was unfindable.

La Nature, 1879-1-166, quotes the *Courrier des Ardennes* as to an occurrence in the Commune Signy-le-Pettit, Easter Sunday, 1879—a conspicuous, isolated house—suddenly its slate roof shot into the air, and then fell to the ground. There had not been a trace of wind. The writer of the account says that the force, which he calls a *trouble inoui* had so singled out this house that nothing in its surroundings beyond a distance of thirty feet had been disturbed.

Scientific American, July 10, 1880—that, according to the Plain-dealer, of East Kent, Ontario, two citizens of East Kent were in a field, and heard a loud report. They saw stones shooting upward from a field. They examined the spot, which was about 16 feet in diameter, finding nothing to suggest an explanation of the occurrence. It is said that there had been neither a whirlwind nor anything else by which to explain.

It may be that witnesses have seen human beings dragged from our own existence either into the objectionable fourth dimension, perhaps then sifting into the fifth, or up to the sky by some exploring thing. I have data, but they are from the records of psychic research. For instance, a man has been seen walking along a road—sudden disappearance. Explanation—that he was not a living human being, but an apparition that had disappeared. I have not been able to develop such data, finding, for instance, that someone in the neighborhood had been reported missing; but it may be that we can find material in our own field.

Upon Dec. 10, 1881, Walter Powell and two companions ascended from Bath in the Government balloon *Saladin* (Valentine and Tomlinson, *Travels in Space*, p. 227). The balloon descended at Bridport, coast of the English Channel. Two of the aëronauts got out, but the balloon, with Powell in it, shot upward. There was a report that the balloon had been seen to fall in the English Channel, near Bridport, but according to Capt. Temple, one of Powell's companions, probably something thrown from the balloon had been seen to fall.

A balloon is lost near or over the sea. If it should fall into the sea it would probably float and for considerable time be a conspicuous object; nevertheless the disappearance of a balloon last seen over the English Channel, cannot, without other circumstances, be considered very mysterious. Now one expects to learn of reports from many places of supposed balloons that had been seen.

But the extraordinary circumstance is that reports came in upon a luminous object that was seen in the sky at the time that this balloon disappeared. In the London *Times*, it is said that a luminous object had been seen, evening of the 13th, moving in various directions in the sky near Cherbourg. It is said that upon the night of the 16th three customhouse guards, at Laredo, Spain, had seen something like a balloon in the sky, and had climbed a mountain in order to see it better, but that it had shot out sparks and had disappeared—and had been reported from Bilbao, Spain, the next day. In the *Morning Post*, it is said that this luminous display was the chief feature; that it was this sparkling that had made the object visible. In the *Standard*, December 16, is an account of something that was seen in the sky, five o'clock, morning of December 15, by Capt. McBain, of the steamship *Countess of Aberdeen*, off the coast of Scotland, 25 miles from Montrose. Through glasses, the object seemed to be a light attached to something thought to be the car of a balloon, increasing and decreasing in size—a large light—"as large as the light at Girdleness." It moved in a direction opposite to that of the wind, though possibly with wind of an upper stratum. It was visible half an hour, and when it finally disappeared, was moving toward Bervie, a town on the Scottish coast about 12 miles north of Montrose. In the *Morning Post* it is said that the explanation is simple: that someone in Monfreith, 8 miles from Dundee, had, late in the evening of the 15th, sent up a fire-balloon, "which had been carried along the coast by a gentle breeze, and, after burning all night, extinguished and collapsed off Montrose, early on Thursday morning (16th)." This story of a balloon that wafted to Montrose, and that was evidently traced until it collapsed near Montrose does not so simply explain an object that was seen 25 miles from Montrose. In the *Standard*, December 19, it is said that two bright lights were seen over Dartmouth Harbor, upon the 11th.

Walter Powell was Member of Parliament for Malmesbury, and had many friends, some of whom started immediately to search. His relatives offered a reward. A steamboat searched the Channel, and did not give up until the 13th; fishing vessels kept on searching. A "sweeping expedition" was organized, and the coast guard was doubled, searching the shore for wreckage, but not a fragment of the balloon, nor from the balloon, except a thermometer in a bag, was found.

In *L'Astronomie*, 1886-312, Prof. Paroisse, of the College Bar-sur-Aube, quotes two witnesses of a *curieux phénomène* that occurred in a garden of the College, May 22, 1886—cloudless sky; wind *tres faible*. Within a small circle in the garden were some: baskets and ashes and a window frame that weighed sixty kilograms. These things suddenly rose from the ground. At a height of about forty feet, they remained suspended several minutes, then falling back to the place from which they had risen. Not a thing outside this small circle had been touched by the seizure. The witnesses said that they had felt no disturbance in the air..

Scientific American, 56-65—that in June, 1886, according to the London *Times,* "a well-known official" was entering Pall Mall,, when he felt a violent blow on the shoulder and heard a hissing, sound. There was no one in sight except a distant policeman. At home, he found that the nap of his coat looked as if a hot wire had been pressed against the cloth, in a long, straight line. No. missile was found, but it was thought that something of a meteoritic nature had struck him.

Charleston News and Courier, Nov. 25, 1886—that, at Edina,, Mo., November 23, a man and his three sons were pulling corn on a farm. Nothing is said of meteorologic conditions, and, for all I know, they may have been pulling corn in a violent thunder storm. Something that is said to have been lightning flashed from the sky. The man was slightly injured, one son killed, the other seriously injured—the third had disappeared. "What has become of him is not known, but it is supposed that he was blinded or crazed by the shock, and wandered away."

Brooklyn Eagle, March 17, 1891—that, at Wilkes-Barre, Pa., March 16th, two men were "lifted bodily and carried considerable distance in a whirlwind." It was a powerful force, but nothing else was affected by it. Upon the same day, there was an occurrence in Brooklyn. In the *New York Times,* March 17, 1891, it is said that two men, Smith Morehouse, of Orange Co., N. Y., and William Owen, of Sussex Co., N. J., were walking in Vanderbilt Avenue, Brooklyn, about 2 o'clock, afternoon of the 16th, when a terrific explosion occurred close to the head of Morehouse, injuring him and stunning Owen, the flash momentarily blinding both. Morehouse's face was covered with marks like powder-marks, and his tongue was pierced. With no one else to accuse, the police arrested Owen, but held him upon the technical charge of intoxication. Morehouse was taken to a hospital, where a splinter of metal, considered either brass or copper, but not a fragment of a cartridge, was removed from his tongue. No other material could be found, though an object of considerable size had exploded. Morehouse's hat had been perforated in six places by unfindable substances. According to witnesses there had been no one within a hundred feet of the men. One witness had seen the flash before the explosion, but could not say whether it had been from something falling or not. In the *Brooklyn Eagle,* March 17, 1891, it is said that neither of the men had a weapon of any kind, and that there had been no disagreement between them. According to a witness, they had been under observation at the time of the explosion, her attention having been attracted by their rustic appearance.

There is an interesting merging here of the findable and the unfindable. I suppose that no one will suppose that someone threw a bomb at these men. But enough substance was found to exclude the notion of "lightning from a clear sky." Something of a meteoritic nature seems excluded.

28

Oᴜᴛ from a round, red planet, a little white shaft—a fairy's arrow shot into an apple. June 10, 1892—a light like a little searchlight, projecting from the limb of Mars. Upon July 11 and 13, it was seen again, by Campbell and Hussey (*Nature*, 50-500).

Aug. 3, 1892—climacteric opposition of Mars.

Upon Aug. 12, 1892, flashes were seen by many persons, in the sky of England. See *Eng. Mec.*, vol. 56. At Manchester, so like signals were they, or so unlike anything commonly known as "auroral" were they, that Albert Buss mistook them for flashes from a lighthouse. They were seen at Dewsbury; described by a correspondent to the *English Mechanic*, who wrote: "I have never seen such an appearance of an aurora." "Rapid flashes" reported from Loughborough.

A shining triangle in a dark circle.

In *L'Astronomie*, 1888-75, Dr. Klein publishes an account of de Speissen's observation of Nov. 23, 1887—a luminous triangle on the floor of Plato. Dr. Klein says it was an effect of sunlight.

In this period, there were in cities of the United States, some of the most astonishing effects at night, in the history of this earth. If Rigel should run for the Presidency of Orion, and if the stars in the great nebula should start to march, there would be a spectacle like those that Grover Cleveland called forth in the United States, in this period.

So then—at least conceivably—something similar upon the moon. Flakes of light moving toward Plato, this night of Nov. 23, 1887, from all the other craters of the moon; a blizzard of shining points gathering into light-drifts in Plato; then the denizens of Aristarchus and of Kepler, and dwellers from the lunar Alps, each raising his torch, marching upon a triangular path, making the triangle shine in the dark—conceivably. Other formations have been seen in Plato, but, according to my records, this symbol that shone in the dark had never been seen before, and has not been seen since.

About two years later—a demonstration of a more exclusive kind— assemblage of all the undertakers of the moon. They stood in a circular formation, surrounded by virgins in their nightgowns—and in nightgowns as nightgowns should be. An appearance in Plinius, Sept. 13, 1889, was reported by Prof. Thury, of Geneva—a black spot with an "intensely white" border.

March 30, 1889—a black spot that was seen for the first time, by Gaudibert, near the center of Copernicus (*L'Astro.*, 1890-235). May 11, 1889—an object as black as ink upon a rampart of Gassendi (*L'Astro.*, 1889-275). It had never been reported before; at the time of the next lunation, it was not seen again. March 30, 1889—a new black spot in Plinius (*L'Astro.*, 1890-187).

The star-like light of Aristarchus—it is a long time since latest preceding appearance (May 7, 1867). Then it cannot be attributed to commonplace lunar circumstances. The light was seen Nov. 7, 1891, by M. d'Adjuda, of the Observatory of Lisbon—"a very distinct, luminous point" (*L'Astro.*, 11-33)

Upon April 1, 1893, a shaft of light was seen projecting from the moon, by M. de Moraes, in the Azores. A similar appearance was seen, Sept. 25, 1893, at Paris, by M. Gaboreau (*L'Astro.*, 13-34).

Another association like that of 1884—in the *English Mechanic*, 55-310, a correspondent writes that, upon May 6, 1892, he saw a shining point (not polar) upon Venus. Upon the 13th of August, 1892, the same object—conceivably—was seen at a short distance from Venus—an unknown, luminous object, like a star of the 7th magnitude that was seen close to Venus, by Prof. Barnard (*Ast. Nach.*, no. 4106).

Upon Aug. 24, 1895, in the period of primary maximum brilliance of Venus, a luminous object, it is said, was seen in the sky, in day time, by someone in Donegal, Ireland. Upon this day, according to the *Scientific American*, 73-374, a boy, Robert Alcorn, saw a large luminous object falling from the sky. It exploded near him. The boy's experience was like Smith Morehouse's. He put his hands over his face: there was a second explosion, shattering his fingers. According to Prof. George M. Minchin no substance of the object that had exploded could be found. Whether there be relation or not, something was seen in the sky of England a week later. In the London *Times*, Sept. 4, 1895, Dr. J. A. H. Murray writes that, at Oxford, a few minutes before 8 P.M., Aug. 31, 1895, he saw in the sky a luminous object, considerably larger than Venus at greater brilliance, emerge from behind tree tops, and sail slowly eastward. It moved as if driven in a strong wind, and disappeared behind other trees. "The fact that it so perceptibly grew fainter as it receded seems to imply that it was not at a great elevation, and so favors a terrestrial origin, though I am unable to conceive how anything artificial. could have presented the same appearance." In the*Times*, of the 6th, someone who had read Dr. Murray's letter says that, about the same time, same evening, he, in London, had seen the same object moving eastward so slowly that he had thought it might be a fire-balloon from a neighboring park. Another correspondent, who had not read Dr. Murray's letter, his own dated September 3, writes from a place not stated that about 8:20 P.M., August 31, he had seen a star-like object, moving eastward, remaining in sight four or five minutes. Then someone who, about 8 P.M., same evening, while driving to the Scarborough station, had seen "a large shooting star," astonishing him, because of its leisurely rate, so different from the velocity of the ordinary "shooting star." There are two other accounts of objects that were seen in the sky, at Bath and at Ramsgate, but not about this time, and I have looked them up in local newspapers, finding that they were probably meteors.

In the *Oxford Times*, September 7, Dr. Murray's letter to the London *Times* is reprinted, with this comment—"We would suggest to the learned doctor that the supposed meteor was one of the fire-balloons let off with the allotments show."

Let it be that when allotments are shown, balloons are always sent up, and that this Editor did not merely have a notion to this effect. Our data are concerned with an object that was seen, at about the same time, at Oxford, about 50 miles southeast of Oxford, and about 170 miles northeast of Oxford, with a fourth observation that we cannot place.

And, in broader terms, our data are concerned with a general expression that objects like ships have been seen to sail close to this earth at times when the planet Venus is nearest this earth. Sept. 18, 1895—inferior conjunction of Venus.

Still in the same period, there were, in London, two occurrences perhaps like that at Donegal. London *Morning Post*, Nov. 16, 1895—that, at noon, November 15, an "alarming explosion" occurred somewhere near Fenchurch Street, London. No damage was done; no trace could be found of anything that had exploded. An hour later, near the Mansion House, which is not far from Fenchurch Street, occurred a still more violent explosion. The streets filled with persons who had run from buildings, and there was investigation, but not a trace could be found of anything that had exploded. It is said that somebody saw "something falling." However, the deadly explainers, usually astronomers, but this time policemen, haunt or arrest us. In the Daily News, though it is not said that a trace of anything that had exploded had been found, it is said that the explanation by the police was that somebody had mischievously placed in the streets fog-signals, which had been exploded by passing vehicles.

Observation by Müller, of Nymegen, Holland—an unknown luminous object that, about three weeks later, was seen near Venus (*Monthly Notices, R. A. S.,* 52-276).

Upon the 28th of April, 1897, Venus was in inferior conjunction. In *Popular Astronomy*, 5-55, it is said that many persons had written to the Editor, telling of "airships" that had been seen, about this time. The Editor writes that some of the observations were probably upon the planet Venus, but that others probably related to toy balloons, "which were provided with various colored lights."

The first group of our data, I take from dispatches to the *New York Sun*, April 2, 11, 16, 18. First of April—"the mysterious light" in the sky of Kansas City —something like a powerful searchlight. "It was directed toward the earth, traveling east at a rate of sixty miles an hour." About a week later, something was seen in Chicago. "Chicago's alleged airship is believed to be a myth, in spite of the fact that a great many persons say that they have seen the mysterious night-wanderer. A crowd gazed at strange lights, from the top of a downtown skyscraper, and Evanston students declare they saw the swaying red and green lights." April 16—reported from Benton, Texas, but this time as a dark object that passed across the moon. Reports from other towns in Texas: Fort Worth, Dallas, Marshall, Ennis, and Beaumont—"It was shaped like a Mexican cigar, large in the middle, and small at both ends, with great wings, resembling those of an enormous butterfly. It was brilliantly illuminated by the rays of two great searchlights, and was sailing in a southeasterly direction, with the velocity of the wind, presenting a magnificent appearance."

New York Herald, April 11—that, at Chicago, night of April 9-10, "until two o'clock in the morning, thousands of amazed spectators declared that the lights seen in the northwest were those of an airship, or some floating object, miles above the earth.... Some declare they saw two cigar-shaped objects and great wings." It is said that a white light, a red light, and a green light had been seen.

There does seem to be an association between this object and the planet Venus, which upon this night was less than three weeks from nearest approach to this earth. Nevertheless this object could not have been Venus, which had set hours earlier. Prof. Hough, of the Northwestern University, is quoted—that the people had mistaken the star *Alpha Orionis* for an airship. Prof. Hough explains that astronomeric effects may have given a changing red and green appearance to this star. *Alpha Orionis* as a northern star is some more astronomy by the astronomers who teach astronomy daytimes and then relax when night comes. That atmospheric conditions could pick out this one star and not affect other brilliant stars in Orion is more astronomy. At any rate the standardized explanation that the thing was Venus disappears.

There were other explainers—someone who said that he knew of an airship (terrestrial one) that had sailed from San Francisco; and had reached Chicago.

Herald, April 12—said that the object had been photographed in Chicago: "a cigar-shaped, silken bag," with a framework—other explanations and identifications, not one of them applying to this object, if it be accepted that it was seen in places as far apart as Illinois and Texas. It is said that, upon March 29th, the thing had been seen in Omaha, as a bright light sailing to the northwest, and that, for a few moments, upon the following night, it had been seen in Denver. It is said that, upon the night of the 9th, despatches had bombarded the newspaper offices of Chicago, from many places in Illinois, Indiana, Missouri, Iowa, and Wisconsin.

"Prof. George Hough maintains that the object seen is *Alpha Orionis*."

April 14—story, veritable observation, yarn, hoax—despatch from Carlensville, Ill.—that upon the afternoon of the 10th, the airship had alighted upon a farm, but had sailed away when approached—"cigar-shaped, with wings, and a canopy on top."

April 15—shower of telegrams—development of jokers and explainers—thing identified as an airship invented by someone in Dodge City, Kansas; identified as an airship invented by someone in Brule, Wisconsin—stories of letters found on farms, purporting to have been dropped by the unknown aeronauts (terrestrial ones)—jokers in various towns, sending up balloons with lights attached—one laborious joker who rigged up something that looked like an airship and put it in a vacant lot and told that it had fallen there—yarn or observation, upon a "queer-looking boat" that had been seen to rise from the water in Lake Erie—continued reports upon a moving object in the sky, and its red and green lights.

Against such an alliance as this, between the jokers and the astronomers, I see small chance for our data. The chance is in the future. If, in April, 1897, extra-mundane voyagers did visit this earth, likely enough they will visit again, and then the alliance against the data may be guarded against.

New York Herald, April 20—that, upon the 19th, about 9 P.M., at Sistersville, W. Va., a luminous object had approached the town from the northwest, flashing brilliant red, white, and green lights.

"An examination with strong glasses left an impression of a huge cone-shaped arrangement 180 feet long, with large fins on either side."

My own general impression:

Night of Oct. 12, 1492—if I have that right. Some night in October, 1492, and savages upon an island-beach are gazing out at lights that they had never seen before. The indications are that voyagers from some other world are nearby. But the wise men explain. One of the most nearly sure expressions in this book is upon how they explain. They explain in terms of the familiar. For instance, after all that is spiritual in a fish passes away, the rest of him begins to shine nights. So there are three big, old, dead things out in the water—

29

THERE have been published several observations upon a signal-like regularity of the Barisal Guns, which, because unaccompanied by phenomena that could be considered seismic, may have been detonations in the sky, and which, because, according to some hearers, they seemed to come from the sky, may have come from some region stationary in the local sky of Barisal. In *Nature*, 61-127, appears a report by Henry S. Schurr, who investigated the sounds in the years 1890-91:

"These Guns are always heard in triplets, i.e., three guns are always heard, one after the other, at regular intervals, and, though several guns may be heard, the number is always three or a multiple of three. Then the interval between the three is always constant, i.e., the interval between the first and the second is the same as the interval between the second and the third, and this interval is usually three seconds, though I have heard it up to ten seconds. The interval, however, between the triplets varies, and varies largely, from a few seconds up to hours and days. Sometimes only one series of triplets is heard in a day; at others the triplets follow with great regularity, and I have counted as many as forty-five of them, one after the other, without pause."

In vols. 16 and 17, *Ciel et Terre*, M. Van den Broeck published a series of papers upon the mysterious sounds that had been heard in Belgium.

July, 1892—heard near Brée, by Dr. Raemaekers, of Antwerp—detonations at regular intervals of about 12 seconds, repeated about 20 times.

Aug. 5, 1892—near Dunkirk, by Prof. Gérard, of Brussels—four reports like sounds of cannons.

Aug. 17, 1893—between Ostend and Ramsgate, by Prof. Gérard—a series of distinct explosions—state of the sky giving no reason to think that they were meteorological manifestations.

Sept. 5, 1893—at Middelkirke—loud sounds of remarkable intensity.

Sept. 8, 1893—English Channel near Dover—by Prof. Gérard—an explosive sound.

In *Ciel et Terre*, 16-485, M. Van den Broeck records an experience of his own. Upon June 25, 1894, at Louvain, he had heard detonations like discharges of artillery: he tabulates the intervals in a series of sounds. If there were signaling from some unknown region over Belgium, and not far from the surface of this earth, or from extra-mundane vessels, and if there were something of the code-like, resembling the Morse alphabet, perhaps, in this series of sounds, there can be small hope of interpreting such limited material, but there may be suggestion to someone to record all sounds and their intervals and modulations, if, with greater duration, such phenomena should ever occur again. The intervals were four minutes and twenty-three minutes; then three minutes, four, three quarters, three and three quarters, three quarters.

Sept. 16, 1895—a triplet of detonations, heard by M. de Schryvere, of Brussels.

There were attempts to explain. Some of M. Van den Broeck's correspondents thought that there had been firing from forts on the coast of England, and somebody thought that the phenomena should be attributed to gravitational effects of the moon. Upon Sept. 13, 1895, four shocks were felt and sounds heard at Southampton: a series of three and then another (*Nature*, 52-552); but I have no other notes upon sounds that were heard in England at this time, except the two explosions that were explained by the police of London. However, M. Van den Broeck says that Mr. Harmer, of Aldeburgh, Suffolk, had, about the first of November, heard booming sounds that had been attributed to cannonading at Harwich. Mr. Harmer had heard other sounds that had been attributed to cannonading somewhere else. He could not offer a definite opinion upon the first sounds, but had investigated the others, learning that the attribution was a mistake.

It was M. de Schryvere's opinion that the triplet of detonations that he had heard was from vessels in the North Sea. But now, according to developments, the sounds of Belgium cannot very well be attributed to terrestrial cannonading in or near Belgium: in *Ciel et Terre*, 16-614, are quoted two artillery officers who had heard the sounds, but could not so trace them: one of these officers had heard a series of detonations with intervals of about two minutes. A variety of explanations was attempted, but in conventional terms, and if these localized, repeating sounds did come from the sky, there's nothing to it but a new variety of attempted explanations, and in most unconventional terms. There are recorded definite impressions that the sounds were in the sky: Prof. Peleseneer's *positivement aérien*. In *Ciel et Terre*, 17-14, M. Van den Broeck announced that General Hennequin, of Brussels, had co-operated with him, and had sent enquiries to army officers and other persons, receiving thirty replies. Some of these correspondents had heard detonations at regular intervals. It is said that the sounds were like cannonading, but not in one instance were the sounds traced to terrestrial gunfire.

Jan. 24, 1896—a triplet of triplets—between 2:30 and 3:30, P.M.—by M. Overloop, of Middelkirke, Belgium—three series of detonations, each of three sounds.

The sounds went on, but, after this occurrence, there seems to me to be little inducement to me to continue upon the subject. This is indication that from somewhere there has been signaling: from extra-mundane vessels to one another, or from some unknown region to this earth, as nearly final as we can hope to find. There are persons who will see nothing but a susceptibility to the mysticism of numbers in a feeling that there is significance in threes of threes. But, if there be attempt in some other world to attract attention upon this earth, it would have to be addressed to some kind of a state of mind that would feel significances. Let our three threes be as mystic as the eleven horns on Daniel's fourth animal; if throughout nature like human nature there be only superstition as to such serialization, that superstition, for want of something more nearly intelligent, would be a susceptibility to which to appeal, and from which response might be expected. I think that a sense of mystic significance in the number three may be universal, because upon this earth it is general, appearing in theologies, in the balanced compositions of all the arts, in logical demonstrations, and in the indefinite feelings that are supposed to be superstitious.

The sounds went on, as if there were experiments, or attempts to communicate by means of other regularizations and repetitions. Feb. 18, 1896—a series of more than 20 detonations, at intervals of 2 or 3 minutes, heard at Ostend, by M. Pulzeys, an engineer of Brussels. Four or five sounds were heard at Ostend by someone else: repeated upon the 21st of February. Heard by M. Overloop, at Ostend, April 6: detonations at 11:57:30 A.M., and at 12:1:32 P.M. Heard the next day, by M. Overloop, at Blankenberghe, at 2:35 and 2:51 P.M.

The last occurrence recorded by M. Van den Broeck was upon the English Channel, May 23, 1896: detonations at 3:20 and 3:40 P.M. I have no more data, as to this period, myself, but I have notes upon similar sounds, by no means so widely reported and commented upon, in France and Belgium about 15 years later. One notices that the old earthquake-explanation as to these sounds has not appeared.

But there were other phenomena in England, in this period, and to considerable degree they were conventionally explained. They were not of the type of the Belgian phenomena, and, because manifestations were seen and felt, as well as heard, they were explained in terms of meteors and earthquakes. But in this double explanation, we meet a divided opposition, and no longer are we held back by the uncompromising attempt by exclusionist science to attribute all disturbances of this earth's surface to a subterranean origin. The admission by Symons and Fordham that we have recorded, as to occurrences of 1887-89, has survived.

The earliest of the accounts that I have read of the quakes in the general region of Worcester and Hereford (London Triangle) that associated with appearances in the sky, was published by two church wardens in the year 1661, as to occurrences of October, 1661, and is entitled, *A True and Perfect Relation of the Terrible Earthquake*. It is said that monstrous flaming things were seen in the sky, and that phenomena below were interesting. We are told, "truly and perfectly," that Mrs. Margaret Petmore fell in labor and brought forth three male offsprings all of whom had teeth and spoke at birth. Inasmuch as it is not recorded what the infants said, and whether in plain English or not, it is not so much an extraordinary birth such as, in one way or another, occurs from time to time, that affronts our conventional notions, as it is the idea that there could be relation between the abnormal in obstetrics and the unusual in terrestrics. The conventional scientist has just this reluctance toward considering shocks of this earth and phenomena in the sky at the same time. If he could accept with us that there often has been relation, the seeming discord would turn into a commonplace, but with us he would never again want to hear of extraordinary detonating meteors exploding only by coincidence over a part of this earth where an earthquake was occurring, or of concussions of this earth, time after time, in one small region, from meteors that, only by coincidence, happened to explode in one little local sky, time after time. Give up the idea that this earth moves, however, and coincidences many times repeated do not have to be lugged in.

Our subject now is the supposed earthquake centering around Worcester and Hereford, Dec. 17, 1896; but there may have been related events, leading up to this climax, signifying long duration of something in the sky that occasionally manifested relatively to this corner of the London Triangle. Mrs. Margaret Petmore was too sensational a person for our liking, at least in our colder and more nearly scientific moments, so we shall not date so far back as the time of her performance; but the so-called earthquakes of Oct. 6, 1863, and of Oct. 30, 1868, were in this region, and we had data for thinking that they were said to be earthquakes only because they could not be traced to terrestrial explosions.

At 5:45 P.M., Nov. 2, 1893, a loud sound was heard at a place ten miles northeast of Worcester, and no shock was felt (*Nature*, 49-245); however at Worcester and in various parts of the west of England and in Wales a shock was felt.

According to James G. Wood, writing in *Symons' Met. Mag.*, 29-8, at 9:30 P.M., Jan. 25, 1894, at Llanthomas and Clifford, towns less than 20 miles west of Hereford, a brilliant light was seen in the sky, an explosion was heard, and a quake was felt. Half an hour later, something else occurred: according to Denning (*Nature*, 49-325) it was in several places, near Hereford and Worcester, supposed to be an earthquake. But, at Stokesay Vicarage, Shropshire (*Symons' Met. Mag.*, 29-8) was seen the same kind of an appearance as that which had been seen at Llanthomas and Clifford, half an hour before: an illumination so brilliant that for half a minute everything was almost as visible as by daylight.

In the *English Mechanic*, 74-155, David Packer calls attention to "a strange meteoric light" that was seen in the sky, at Worcester, during the quake of Dec. 17, 1896. I should say that this was the severest shock felt in the British Isles, in the 19th century, with the exception of the shock of April 22, 1884, in the eastern point of the London Triangle. There was something in the sky. In *Nature*, 55-179, J. Lloyd Bozward writes that, at Worcester, a great light was seen in the sky, at the time of the shock, and that, in another town, "a great blaze" had been seen in the sky. In *Symons' Met. Mag.*, 31-180, are recorded many observations upon lights that were seen in the sky. In an appendix to his book, *The Hereford Earthquake of 1896*, Dr. Charles Davison says that at the time of the quake (5:30 A.M.) there was a luminous object in the sky, and that it "traversed a large part of the disturbed area." He says that it was a meteor, and an extraordinary meteor that lighted up the ground so that one could have picked up a pin. With the data so far considered, almost anyone would think that of course an object had exploded in the sky, shaking the earth underneath. Dr. Davison does not say this. He says that the meteor only happened to appear over a part of this earth where an earthquake was occurring, "by a strange coincidence."

Suppose that, with ordinary common sense, he had not lugged in his "strange coincidence," and had written that of course the shock was concussion from an explosion in the sky

Shocks that had been felt before midnight, December 17, and at 1:30 or 1:45, 2, 3, 3:30, 4, 5, and 5:20, and then others at 5:40 or 5:45 and at 6:15 o'clock— and were they, too, concussions, but fainter and from remoter explosions in the sky—and why not, if of course the great shock of 5:30 o'clock was from a great explosion in the sky—and by what multiplication of strangeness of coincidence could detonating meteors, or explosions of any other kind, so localize in the one little sky of Worcester, if this earth be a moving earth—and how could their origin be otherwise than a fixed region nearby?

In some minds it may be questionable that the earth could be so affected as it was at 5:30 A.M., Dec. 17, 1896, by an explosion in the sky. Upon Feb. 10, 1896, a tremendous explosion occurred in the sky of Madrid: throughout the city windows were smashed; a wall in the building occupied by the American

Legation was thrown down. The people of Madrid rushed to the streets, and there was a panic in which many were injured. For five hours and a half a luminous cloud of débris hung over Madrid, and stones fell from the sky.

Suppose, just at present, we disregard all the Worcester-Hereford phenomena except those of Dec. 17, 1896. Draw a diagram, illustrating a stream of meteors pursuing this earth, now supposed to be rotating and revolving, for more than 400,000 miles in its orbit, and curving around gracefully and unerringly after the rotating earth, so as to explode precisely in this one little local sky and nowhere else. But we can't think very reasonably even of a flock of birds flying after and so precisely pecking one spot on an apple thrown in the air by somebody. Another diagram—stationary earth—bombardment of any kind one chooses to think of—same point hit every time—thinkable.

The phenomena associate with an opposition of Mars. Dec. 10, 1896—opposition of Mars.

But we have gone on rather elaborately with perhaps an insufficiency to base upon. We cannot say, directly, that all the phenomena of the night of Dec. 16-17, 1896, were shocks from explosions in the sky: only during the greatest of the concussions was something seen, or was something near enough to be seen.

We apply the idea of the diagrams to another series of occurrences in this period. Now draw a diagram relatively to the sky of Florida, and see just what the explanation of coincidence demands or exacts. But then consider the diagram as one of an earth that does not move and of something that is fixed over a point upon its surface. Things can be thought of as coming down from somewhere else to one special sky of this earth, as logically as precariously placed objects on one special window sill sometimes come down to a special neighbor.

In the *Monthly Weather Review*, 23-57, is a report, by the Director of the Florida Weather Service, upon "mysterious sounds" and luminous effects in the sky of Florida. According to investigation, these phenomena did occur in the sky of Florida, about noon, Feb. 7, 1895, again at 5 o'clock in the morning of the 8th, and again between 6 and to o'clock, night of the 8th. The Editor of the *Review* thinks that three meteors may have exploded so in succession in the sky of Florida, and nowhere else, "by coincidence."

30

CHAR me the trunk of a redwood tree. Give me pages of white chalk cliffs to write upon. Magnify me thousands of times, and replace my trifling immodesties with a titanic megalomania—then might I write largely enough for our subjects. Because of accessibility and abundance of data, our accounts deal very much with the relatively insignficant phenomena of Great Britain. But our subject, if not so restricted, would be the violences that have screamed from the heavens, lapping up villages with tongues of fire. If, because of appearances in the sky, it be accepted that some of the so-called earthquakes of Italy and South America represented relations with regions beyond this earth, then it is accepted that some of this earth's greatest catastrophes have been relations with the unknown and the external. We have data that seem to be indications of signaling, but not unless we can think that foreign giants have hurled explosive mountains at this earth can we see such indications in all the data.

Our data do seem to fall into two orders of phenomena: sounds of Melida, Barisal, and Belgium, and nothing falling from the sky, and nothing seen in the sky, and excellently supported observations for accepting a signal-like intent in intervals and grouping of sounds, at least in Barisal and Belgium; and the unregularized phenomena of Worcester-Hereford, Colchester, Comrie, and Birmingham, in which appearances are seen in the sky, or in which substances fall from the sky, and in which effects upon this earth, not noted at all in Belgium and Bengal, are great, and sometimes tremendous. It seems that extra-geography divides into the extra-sociologic and the extra-physical; and in the second type of phenomena, we suppose the data are of physical relations between this earth and other worlds. We think of a difference of potential. There were tremendous detonations in the sky at the times of the falls of the little black stones of Birmingham and Wolverhampton, and the electric manifestations, according to descriptions in the newspapers, were extraordinary, and great volumes of water fell. Consequently the events were supposed to be thunderstorms. I suppose, myself, that they were electric storms, but electric storms that represented difference of potential between this earth and some region that was fixed, at least eleven years, over Birmingham and Wolverhampton, bringing down stones and volumes of water from some other world, or bringing down stones, and dislodging intervening volumes of water, such as we have many data for thinking exist in outer space, sometimes in bodies of warm or hot water, and sometimes as great masses, or fields; of ice.

Let two objects be generically similar, but specifically different and a relation that may be known as a difference of potential, though that term is usually confined to electric relations, generates between them. Quite as the Gulf Stream—though there are no reasons to suppose that there is such a Gulf Stream as one reads of—represents a relation between bodies of water heated differently, given any two worlds, alike in general constitution, but differing, say, electrically, and given proximity, we conceive of relations between them other than gravitational.

But this cloistered earth, and its monkish science—shrinking from, denying, or disregarding, all data of external relations, except some one controlling force that was once upon a time known as Jehovah, but that has been re-named Gravitation—

That the electric exchanges that were recognized by the ancients, but that were anthropomorphically explained by them, have poured from the sky and have gushed to the sky, afferently and efferently, between this earth and the nearby planets, or between this mainland and its San Salvadors, and have been recognized by the moderns, or the neo-ancients, but have been meteorologically and seismologically misconstrued by them.

When a village spouts to the sky, it is said to have been caught up in a cyclone: when unknown substances fall from the sky, not much of anything is said upon the subject.

Lost tribes and the nations that have disappeared from the face of this earth— that the skies have reeked with terrestrial civilizations, spreading out in celestial stagnations, where their remains to this day may be. The Mayans— and what became of them? Bones of the Mayans, picked white as frost by space-scavengers, regioned to this day in a sterile luxuriance somewhere, spread upon existence like the pseudo-breath of Death, crystallized on a sky-pane. Three times gaps wide and dark the history of Egypt—and that these abysses were gulfed by disappearances—that some of the eliminations from this earth may have been upward translations in functional suctions. We conceive of Supervision upon this earth's development, but for it the names of Jehovah and Allah seem old-fashioned—that the equivalence of wrath, but like the storms of cells that, in an embryonic thing, invade and destroy cartilage-cells, when they have outlived their usefulness, have devastated this earth's undesirables. Likely enough, or not quite likely enough, one of these earlier Egypts was populated by sphinxes, if one can suppose that some of the statuary still extant in Egypt were portraitures. This is good, though also not so good, orthodox Evolutionary doctrine—that between types occur transitionals —

That Elimination and Redistribution swept an earlier Egypt with suctions— because it was written, in symbols of embryonic law, that life upon this earth must form onward—and the crouching sphinx on the sands of Egypt, blinking the mysticism of her morphologic mixtures, would perhaps detain forever the less interesting type that was advancing—

That often has Clarification destroyed transitionals, that they shall not hold back development.

One conceives of their remains, to this day, wafting still in the currents of the sky: floating avenues of frozen sphinxes, solemnly dipping in cosmic undulations, down which circulate processions of Egyptian mummies.

An astronomer upon this earth notes that things in parallel lines have crossed the sun.

We offer this contribution as comparing favorably with the works of any other historian. We think that some of the details may need revision, but that what they typify is somewhere nearly acceptable:

Latitudes and longitudes of bones, not in the sky, but upon the surface of this earth. Baron Toll and other explorers have, upon the surface of this earth, kicked their way through networks of ribs and protrusions of skulls and stacks of vertebræ, as numerous as if from dead land they had sprouted there. Anybody who has read of these tracts of bones upon the northern coast of Siberia, and of some of the outlying islands that are virtually composed of bones cemented with icy sand, will agree with me that there have been cataclysms of which conventionality and standardization tell us nothing. Once upon a time, some unknown force translated, from somewhere, a million animals to Colorado, where their remains now form great bone-quarries. Very largely do we express a reaction against dogmatism, and sometimes we are not dogmatic, ourselves. We don't know very positively whether at times the animal life of some other world has been swept away from that world, eventually pouring from the sky of Siberia and of Colorado, in some of the shockingest floods of mammoths from which spattered cats and rabbits, in cosmic scenery, or not. All that we can say is that when we turn to conventionality it is to blankness or suppression. Every now and then, to this day, occurs an alleged fall of blood from the sky, and I have notes upon at least one instance in which the microscopically examined substance was identified as blood. But now we conceive of intenser times, when every now and then a red cataract hung in the heavens like the bridal veil of the goddess of murder. But the science of today is a soporific like the idealism of Europe before the War broke out. Science and idealism—wings of a vampire that lulls consciousness that might otherwise foresee catastrophe. Showers of frogs and showers of fishes that occur to this day—that they are the dwindled representatives to this day of the cataclysms of intenser times when the skies of this earth were darkened by afferent clouds of dinosaurs. We conceive of intenser times, but we conceive of all times as being rhythmic times. We are too busy to take up alarmism, but, if Rome, for instance, never was destroyed by terrestrial barbarians, if we cannot very well think of Apaches seizing

Chicago, extra-mundane vandals may often have swooped down upon this earth, and they may swoop again; and it may be a comfort to us, some day, to mention in our last gasp that we told about this.

History, geology, palæontology, astronomy, meteorology—that nothing short of cataclysmic thinking can break down these united walls of Exclusionism.

Unknown monsters sometimes appear in the ocean. When, upon the closed system of normal preoccupations, a story of a sea serpent appears, it is inhospitably treated. To us of the wider cordialities, it has recommendations for kinder reception. I think that we shall be noted in recognitions of good works for our bizarre charities. Far back in the topography of the nineteenth century, Richard Proctor was almost submerged in an ocean of smugness, but now and then he was a little island emerging from the gently alternating doubts and satisfactions of his era, and by means of several papers upon the "sea serpent" he so protruded and gave variety to a dreary uniformity. Proctor reviewed some of the stories of "sea serpents." He accepted some of them. This will be news to some conventionalists. But the mystery that he could not solve is their conceivable origin. To be sure this earth may not be round, or top-shaped, and may tower away somewhere, perhaps with the great Antarctic plateau as its foothills, to a gigantic existence commensurate throughout with the sea monsters that sometimes reach regions known to us. Judging by our experience in other fields of research, we suspect that this earth never has been traversed except in conventional trade-routes and standard explorations. One supposes that enormous forms of life that have appeared upon the surface of the ocean, did not come from conditions of great pressure below the surface. If there be no habitat of their own, in unknown seas of this earth, the monsters fell from the sky, surviving for a while. In his day, Charles Lyell never said a more preposterous thing than this—however, we have no idea that mere preposterousness is a criterion.

Then at times the things have fallen upon land, presumably. To scientific minds in their present anæmia of .malnutrition, we offer new nourishment. There are materials for a science of neo-palæontology—as it were—at least a new view of animal-remains upon this earth. Remains of monsters, supposed to have lived geologic ages ago, are sometimes found, not in ancient deposits, but upon, or near, the surface of the ground, sometimes barely covered. I have notes upon a great pile of bones, supposed to be the remains of a whale, out in open view in a western desert.

In the American Museum of Natural History, New York City, is the mummified body of a monster called a *trachodon*, found in Converse County, Wyoming. It was not found upon the surface of the ground, which is bad for our attempts to stimulate palæontology. But the striking datum to me is that the only other huge mummy that I know of is another *trachodon*, now in the Museum of Frankfort. If only extraordinarily would geologic processes mummify remains of a huge animal, doubly extraordinarily would two animals of the same species be so exclusively affected. One at least gives some consideration to the idea that these *trachodons* are not products of geologic circumstances, but were affected, in common, by other circumstances. By inspiration, or progressive deterioration, one then conceives of the things as having wafted and dried in space, finally falling to this earth. Our swooping vandals are relieved with showering mummies. Life is turning out to be interesting.

Organic substances like life-fluids of living things have rained from the sky. However, it is enough for our general purposes to make acceptable simply that unknown substances have, in large quantities, fallen from the sky. That is neo-ism enough, it seems to me. I consider, myself, all such data relatively to this earth's stationariness or possible motions. In *Ciel et Terre*, 22-198, it is said that, about 2 P.m., June 8, 1901, a glue-like substance fell at Sart. The story is told by an investigator, M. Michael, a meteorologist. He says that he saw this substance falling from the sky, but does not give an estimate of duration: he says that he arrived during the last five minutes of the shower. Editors and extra-geographers can't help trying to explain. The Editor of *Ciel et Terre* writes that, three days before, there had been, at Antwerp, a great fire, in which, among other substances, a large quantity of sugar had been burned. He asks whether there could be any connection. Antwerp is about 80 miles from Sart.

Sept. 2, 1905—the tragedy of the space-pig:

In the *English Mechanic*, 86-100, Col. Markwick writes that, according to the *Cambrian Natural Observer*, something was seen in the sky, at Llangollen, Wales, Sept. 2, 1905. It is described as an intensely black object, about two miles above the earth's surface, moving at the rate of about twenty miles an hour. Col. Markwick writes: "Could it have been a balloon?" We give Col. Markwick good rating as an extra-geographer, but of the early, or differentiating type, a transitional, if not a sphinx: so he was not quite developed enough to publish the details of this object. In the *Cambrian Natural Observer*, 1905-35—the journal of the Astronomical Society of Wales —it is said that, according to accounts in the newspapers, an object had appeared in the sky, at Llangollen, Wales, Sept. 2, 1905. At the schoolhouse, in Vroncysylite—I think that's it: with all my credulity, some of these Welsh names look incredible to me, in my notes—the thing in the sky had been examined through powerful field glasses. We are told that it had short wings, and flew, or moved, in a way described as "casually inclining sideways." It seemed to have four legs, and looked to be about ten feet long. According to several witnesses it looked like a huge, winged pig, with webbed feet. "Much speculation was rife as to what the mysterious object could be."

Five days later, according to a member of the Astronomical Society of Wales —see *Cambrian Observer*, 1905-30—a purple-red substance fell from the sky, at Llanelly, Wales.

I don't know that my own attitude toward these data is understood, and I don't know that it matters in the least; also from time to time my own attitude changes: but very largely my feeling is that not much can be, or should be, concluded from our meager accounts, but that so often are these occurrences, in our fields, reported, that several times every year there will be occurrences that one would like to have investigated by someone who believes that we have written nothing but bosh, and by someone who believes in our data almost religiously. It may be that, early in February, 1892, a luminous thing traveled back and forth, exploring for ten hours in the sky of Sweden. The story is copied from a newspaper, and ridiculed, in the *English Mechanic*, 55-34. Upon March 7, 1893, a luminous object shaped like an elongated pear was seen in the sky of Val-de-la-Haye, by M. Raimond Coulon (*L'Astro.*, 1893-169). M. Coulon's suggestion is that the light may have been a signal suspended from a balloon. The signal-idea is interesting.

In the summer of 1897, several weeks after Prof. Andrée and his two companions had sailed in a balloon, from Amsterdam Island, Spitzbergen, it was reported that a balloon had been seen in British Columbia. There was wide publicity: the report was investigated. It may be that had a terrestrial balloon escaped from somewhere in the United States or Canada, or if there had been a balloon-ascension at this time, the circumstances would have been reported: it may well be that the object was not Andrée's balloon. President Bell, of the National Geographic Society, heard of this object, and heard that details had been sent to the Swedish Foreign Office, and cabled to the American Minister, at Stockholm, for information. He publishes his account in the *National Geographic Magazine*, 9-102. He was referred to the Swedish Consul, at San Francisco. In reply to inquiry, the Consul telegraphed the following data, which had been collected by the President of the Geographical Society of the Pacific:

"Statement of a balloon passing over the Horse-Fly Hydraulic Mining Camp, in Caribou, British Columbia, 52°, 20´, and Longitude 120°, 30´—

"From letters of J. B. Robson, manager of the Caribou Mining Co., and of Mrs. Wm. Sullivan, the blacksmith's wife, there, and a statement of Mr. John J. Newsome, San Francisco, then at camp. About 2 or 3 o'clock, in the afternoon, between fourth and seventh of August last, weather calm and cloudless, Mrs. Sullivan, while looking over the Hydraulic Bank, noticed a round, grayish-looking object in the sky, to the right of the sun. As she watched, it grew larger and was descending. She saw the larger mass of the balloon above, and a smaller mass apparently suspended from the larger. It continued to descend, until she plainly recognized it as a balloon and a large basket hanging thereto. It finally commenced to swing violently back and forth, and move very fast toward the eastward and northward. Mrs. Sullivan called her daughter, aged 18, and about this time Mrs. Robson and her daughter were observing it."

If someone saw a strange fish in the ocean, we'd like to know—what was it like? Stripes on him—spots—what? It would be unsatisfactory to be told over and over only that a dark body had crossed some waves. In *Cosmos*, n.s., 39-356, a satisfactory correspondent writes that, at Lille, France, Sept. 4, 1898, he saw a red object in the sky. It was like the planet Mars, but was in the position of no known planet. He looked through his telescope, and saw a rectangular object, with a violent-colored band on one side of it, and the rest of it striped with black and red. He watched it ten minutes, during which time it was stationary; then, like the object that was seen at the time of the Powell-mystery, it cast out sparks and disappeared.

In the *English Mechanic*, 75-417, Col. Markwick writes that, upon May 10, 1902, a friend of his had seen in the sky, in South Devon, a great number of highly colored objects like little suns or toy-balloons. "Altogether beats me," says Col. Markwick.

Upon March 2, 1899, a luminous object in the sky, from 10 A.M., until 4 P.M., was reported from El Paso, Texas. Mentioned in the *Observatory*, 22-247—supposed to have been Venus, even though Venus was then two months past secondary maximum brilliance. This seems reasonable enough, in itself, but there are other data for thinking that an unknown, luminous body was at this time in the especial sky of the southwestern states. In the *U. S. Weather Bureau Report* (*Ariz. Sec.*, March, 1899) it is said that, at Prescott, Arizona, Dr. Warren E. Day had seen a luminous object, upon the 8th of March, "that traveled with the moon" all day, until 2 P.M. It is said that, the day before, this object had been seen close to the moon, by Mr. G. O. Scott, at Tonto, Arizona. Dr. Day and Mr. Scott were voluntary observers for the *Weather Review*. This association with the moon and this localization of observation are puzzling.

La Nature (Sup.) Nov. 11, 1899—that at Luzarches, France, upon the 28th of October, 1899, M. A. Garrie had seen, at 4:50 P.M., a round, luminous object rising above the horizon. About the size of the moon. He watched it for 15 minutes, as it moved away, diminishing to a point. It may be that something from external regions was for several weeks in the especial sky of France. In *La Nature* (*Sup.*) Dec. 16, 1899, someone writes that he had seen, Nov. 15,

1899, 7 P.M., at Dourite (Dordogne) an object like an enormous star, at times white, then red, and sometimes blue, but moving like a kite. It was in the south. He had never seen it before. Someone, in the issue of December 30th, says that without doubt it was the star Formalhaut, and asks for precise position. Issue of Jan. 20, 1900—the first correspondent says that the object was in the southwest, about 35 degrees above the horizon, but moving so that the precise position could not be stated. The kite-like motion may have been merely seeming motion—object may have been Formalhaut, though 35 degrees above the horizon seems to me to be too high for Formalhaut—but, then, like the astronomers, I'm likely at times to expose what I don't know about astronomy. Formalhaut is not an enormous star. Seventeen are larger.

May 1, 1908, between 8 and 9 P.M., at Vittel, France—an object, with a nebulosity around it, diameter equal to the moon's, according to a correspondent to *Cosmos*, n.s., 58-535. At 9 o'clock a black band appeared upon the object, and moved obliquely across it, then disappearing. The Editor thinks that the object was the planet Venus, under extraordinary meteorologic conditions.

Dark obj., by Prof. Brooks, July 21, 1896 (*Eng. Mec.*, 64-12); dark obj., by Gathmann, Aug. 22, 1896 (*Sci. Amer. Sup.*, 67-363); two luminous objs., by Prof. Swift, evidently in a local sky of California, because unseen elsewhere in California, Sept. 20, and one of them again, Sept. 21, 1896 (*Astro. Jour.* 17-8, 103); "Waldemath's second moon," Feb. 5, 1898 (*Eng. Mec.*, 67-545); unknown obj., March 30, 1908 (*Observatory*, 31-215); dark obj., Nov. 10, 1908 (*Bull. Soc. Astro. de France*, 23-74).

31

Cold Harbor, Hanover Co., Virginia—two men in a field—"an apparently clear sky." In the *Monthly Weather Review*, 28-29, it is said that upon Aug. 7, 1900, two men were struck by lightning. The Editor says that the weather map gave no indication of a thunderstorm, nor of rain, in this region at the time.

In July, 1904, a man was killed on the summit of Mt. San Gorgionio, near the Mojave desert. It is said that he was killed by lightning. Two days later, upon the summit of Mt. Whitney, 180 miles away, another man was killed "by lightning" (*Ciel et Terre*, 29-120).

It is said, in *Ciel et Terre*, 17-42, that, in the year 1893, nineteen soldiers were marching near Bourges, France, when they were struck by an unknown force. It is said that in known terms there is no explanation. Some of the men were killed, and others were struck insensible. At the inquest it was testified that there had been no storm, and that nothing had been heard.

If there occur upon the surface of this earth pounces from blankness and seizures by nothings, and "sniping" with bullets of unfindable substance, we nevertheless hesitate to bring witchcraft and demonology into our fields. Our general subject now is the existence of a great deal that may be nearby, or temporarily nearby, ordinarily invisible, but occasionally revealed by special circumstances. A background of stars is not to be compared, in our data, with the sun for a background, as a means of revelations. We accept that there are sunspots, but we gather from general experience and special instances that the word "sunspot" is another of the standardizing terms like "auroral" and "meteoric" and "earthquakes." See Webb's *Celestial Objects*for some observations upon large definite obscurations called "sunspots" but which were as evanescent against the sun as would be islands and jungles of space, if intervening only a few moments between this earth and the swifting moving sun. According to Webb, astronomers have looked at great obscurations upon the sun, have turned away, and then looked again, finding no trace of the phenomena. Eclipses are special circumstances, and rather often have large, unknown bulks been revealed by different light-effects during eclipses. For instance, upon Jan. 22, 1898, Lieut. Blackett, R.N., assisting Sir Norman Lockyer, at Viziadrug, India, during the total eclipse of the sun, saw an unknown body between Venus and Mars (*Jour. Leeds Astro. Soc.*, 1906-23). We have had other instances, and I have notes upon still more. The photographic plate is a special condition, or sensitiveness. In *Knowledge*, 16-234, a correspondent writes that, in August, 1893, in Switzerland, moonlighted night, he had exposed a photographic plate for one hour. Upon the photograph, when developed, were seen irregular, bright markings, but there had been no lightning to this correspondent's perceptions.

The details of the sheep-panic of Nov. 3, 1888, are extraordinary. The region affected was much greater than was supposed by the writer whom we quoted in an earlier chapter. It is said in another account in *Symons' Meteorological Magazine,* that, in a tract of land twenty-five miles long and eight miles wide, thousands of sheep had, by a simultaneous impulse, burst from their bounds; and had been found the next morning, widely scattered, some of them still panting with terror under hedges, and many crowded into corners of fields. See London *Times*, Nov. 20, 1888. An idea of the great number of flocks affected is given by one correspondent who says that malicious mischief was out of the question, because a thousand men could not have frightened and released all these sheep. Someone else tries to explain that, given an alarm in one flock, it might spread to the others. But all the sheep so burst from their folds at about eight o'clock in the evening, and one supposes that many folds were far from contiguous, and one thinks of such contagion requiring considerable time to spread over 200 square miles. Something of an alarming nature and of a pronounced degree occurred somewhere near Reading, Berkshire, upon this evening. Also there seems to be something of special localization: the next year another panic occurred in Berkshire not far from Reading.

I have a datum that looks very much like the revelation of a ghost-moon, though I think of it myself in physical terms of light-effects. In *Country Queries and Notes*, 1-138, 417, it is said that, in the sky of Gosport, Hampshire, night of Sept. 14, 1908, was seen a light that came as if from an unseen moon. It may be that I can here record that there was a moon-like object in the sky of the Midlands and the south of England, this night, and that, though to human eyesight, this world, island of space, whatever it may have been, was invisible, it was, nevertheless, revealed. Upon this evening of Sept. 14, 1908, David Packer, then in Northfield, Worcestershire, saw a luminous appearance that he supposed was auroral, and photographed it. When the photograph was developed, it was seen that the "auroral" light came from a large, moon-like object. A reproduction of the photograph is published in the *English Mechanic*, 88-211. It shows an object as bright and as well-defined as the conventionally accepted moon, but only to the camera had it revealed itself, and Mr. Packer had caught upon a film a space-island that had been invisible to his eyes. It seems so, anyway.

In *Country Queries and Notes*, 1-328, it is said that, upon Aug. 2, 1908, at Ballyconneely, Connemara coast of Ireland, was seen a phantom city of different-sized houses, in different styles of architecture; visible three hours. It is said that no doubt the appearance was a mirage of some city far away—far away, but upon this earth, of course. This apparition is not of the type that we consider so especially of our own data. The so-called mirages that so especially interest us are interesting to us not in themselves, but in that they belong to the one order of phenomena or evidence that unifies so many fields of our data: that is, repetitions in a local sky, signifying the fixed position of something relatively to a small part of this earth's surface. We cannot think that mirages, terrestrial or extraterrestrial, could so repeat. But if in a local sky of this earth there be a fixed region, perhaps not a city, but something of rugged and featureful outlines, with projections that might look architectural, reflections from it, shadows, or Brocken specters repeating always in one special sky are thinkable except by the Chinese-minded who regard all our data as "foreign devils." The writer in *Country Queries and Notes* says —"Circumstantial accounts have even been published of the city of Bristol being distinctly recognized in a mirage seen occasionally in North America." If we shall accept that anywhere in North America repeated representations of the same city or city-like scene have appeared in the same local sky, I prefer, myself, a foreign devil of a thought, and its significance, whether hellish or not, that this earth is stationary, to such a domestic vagrant of a thought as the idea that mirage could so pick out the city of Bristol, or any other city, over and over, and also invariably pick out for its screen the same local sky, thousands of miles, or five miles, away.

In the *English Mechanic*, Sept. 10, 1897, a correspondent to the *Weekly Times and Echo* is quoted. He had just returned from the Yukon. Early in June, 1897, he had seen a city pictured in the sky of Alaska. "Not one of us could form the remotest idea in what part of the world this settlement could be. Some guessed Toronto, others Montreal, and one of us even suggested Pekin. But whether this city exists in some unknown world on the other side of the North Pole, or not, it is a fact that this wonderful mirage occurs from time to time yearly, and we were not the only ones who witnessed the spectacle. Therefore it is evident that it must be the reflection of some place built by the hand of man." According to this correspondent, the "mirage" did not look like one of the cities named, but like "some immense city of the past."

In the *New York Tribune*, Feb. 17, 1901, it is said that Indians of Alaska had told of an occasional appearance, as if of a city, suspended in the sky, and that a prospector, named Willoughby, having heard the stories, had investigated, in the year 1887, and had seen the spectacle. It is said that, having several times attempted to photograph the scene, Willoughby did finally at least show an alleged photograph of an aërial city. In *Alaska*, p. 140, Miner Bruce says that Willoughby, one of the early pioneers in Alaska, after whom Willoughby Island is named, had told him of the phenomenon, and that, early in 1899, he had accompanied Willoughby to the place over which the mirage was said to repeat. It seems that he saw nothing himself, but he quotes a member of the Duc d'Abruzzi's expedition to Mt. St. Elias, summer of 1897, Mr. C. W. Thornton, of Seattle, who saw the spectacle, and wrote—"It required no effort of the imagination to liken it to a city, but was so distinct that it required, instead, faith to believe that it was not in reality a city." Bruce publishes a reproduction of Willoughby's photograph, and says that the city was identified as Bristol, England. So definite, or so un-mirage-like, is this reproduction, trees and many buildings shown in detail, that one supposes that the original was a photograph of a good-sized terrestrial city, perhaps Bristol, England.

In Chapter 10, of his book, *Wonders of Alaska*, Alexander Badlam tries to explain. He publishes a reproduction of Willoughby's photograph: it is the same as Bruce's, except that all buildings are transposed, or are negative in positions. Badlam does not like to accuse Willoughby of fraud: his idea is that some unknown humorist had sold Willoughby a dry plate, picturing part of the city of Bristol. My own idea is that something of this kind did occur, and that this photograph, greatly involved in accounts of the repeating mirages, had nothing to do with the mirages. Badlam then tells of another photograph. He tells that two men, near the Muir Glacier, had, by means of a pan of quicksilver, seen a reflection of an unknown city somewhere, and that their idea was that it was at the bottom of the sea near the glacier, reflecting in the sky, and reflecting back to and from the quicksilver. That's complicated. A photographer named Taber then announced that he had photographed this scene, as reflected in a pan of quicksilver. Badlam publishes a reproduction of Taber's photograph, or alleged photograph. This time, for anybody who prefers to think that there is, somewhere in the sky of Alaska, a great, unknown city, we have a most agreeable photograph: exotic-looking city; a structure like a coliseum, and another prominent building like a mosque, and many indefinite, mirage-like buildings. I'd like to think this photograph genuine, myself, but I do conceive that Taber could have taken it by photographing a panorama that he had painted. Badlam's explanation is that mirages of glaciers are common, in Alaska, and that they look architectural. Some years ago, I read five or six hundred pounds of literature upon the Arctic, and I should say that far-projected mirages are not common in the Arctic: mere looming is common. Badlam publishes a photograph of a mirage of Muir Glacier. The looming points of ice do look Gothic, but they are obviously only loomings, extending only short distances from primaries, with no detachment from primaries, and not reflecting in the sky.

For the first identification of the Willoughby photograph as a photograph of part of the city of Bristol, see the *New York Times*, Oct. 20, 1889. That this photograph was somebody's hoax seems to be acceptable. But it was not similar to the frequently reported scene in the sky of Alaska, according to descriptions. In the *New York Times*, Oct. 31, 1889, is an account, by Mr. L. B. French, of Chicago, of the spectral representation, as he saw it, near Mt. Fairweather. "We could see plainly houses, well-defined streets, and trees. Here and there rose tall spires over huge buildings, which appeared to be ancient mosques or cathedrals…. It did not look like a modern city—more like an ancient European City."

Jour. Roy. Met. Soc., 27-158:

That every year, between June 21 and July 10, a "phantom city" appears in the sky, over a glacier in Alaska; that features of it had been recognized as buildings in the city of Bristol, England, so that the "mirage" was supposed to be a mirage of Bristol. It is said that for generations these repeating representations had been known to the Alaskan Indians, and that, in May, 1901, a scientific expedition from San Francisco would investigate. It is said that, except for slight changes, from year to year, the scene was always the same.

La Nature, 1901-1-303: That a number of scientists had set out from Victoria, B. C., to Mt. Fairweather, Alaska, to study a repeating mirage of a city in the sky, which had been reported by the Duc d'Abruzzi, who had seen it and had sketched it.

32

Nɪɢʜᴛ of Dec 7, 1900—for seventy minutes a fountain of light played upon the planet Mars.

Prof. Pickering—"absolutely inexplicable" (*Sci. Amer.*, 84-179).

It may have been a geyser of messages. It may be translated some day. If it were expressed in imagery befitting the salutation by a planet to its dominant, it may be known some day as the most heroic oration in the literature of this geo-system. See Lowell's account in *Popular Astronomy*, 10-187. Here are published several of the values in a possible code of long flashes and short flashes. Lowell takes a supposed normality for unity, and records variations of two thirds, one and one third, and one and a half. If there be, at Flagstaff, Arizona, records of all the long flashes and short flashes that were seen, for seventy minutes, upon this night of Dec. 7, 1900, it is either that the greetings of an island of space have been hopelessly addressed to a continental stolidity, or there will have to be the descent, upon Flagstaff, Arizona, by all the amateur Champollions of this earth, to concentrate in one deafening buzz of attempted translation.

It was at this time that Tesla announced that he had received, upon his wireless apparatus, vibrations that he attributed to the Martians. They were series of triplets.

It is our expression that, during eclipses and oppositions and other notable celestial events, lunarians try to communicate with this earth, having a notion that at such times the astronomers of this earth may be more nearly alert.

An eclipse of the moon, March 10-11, 1895—not a cloud; no mist—electric flashes like lightning, reported from a ship upon the Atlantic (*Eng. Mec.*, 61-100).

During the eclipse of the sun, July 29, 1897, a strange image was taken on a sensitive plate, by Mr. L. E. Martindale, of St. Mary's, Ohio. It looks like a record of knotted lightning. See *Photography*, May 26, 1898.

In the *Bull. Soc. Astro. de France*, 17-205, 315, 447, it is said that upon the first and the third of March, 1903, a light like a little star, flashing intermittently, was seen by M. Rey, in Marseilles, and by Maurice Gheury, in London, in the lunar crater Aristarchus. March 28, 1903—opposition of Mars.

In *Cosmos*, n.s., 49-259, M. Desmoulins writes, from Argenteuil, that, upon Aug. 9, 1903, at 11 P.M., moving from north to south, he saw a luminous object. The planet Venus was at primary greatest brilliance upon Aug. 13, 1903. In three respects it was like other objects that have been observed upon this earth at times of the nearest approach of Venus: it was a red object; it appeared only in a local sky, and it appeared in the time of the visibility of Venus. With M. Desmoulins were four persons, one of whom had field glasses. The object was watched twenty minutes, during which time it traveled a distance estimated at five or six kilometers. It looked like a light suspended from a balloon, but, through glasses, no outline of a balloon could be seen, and there were no reflections of light as if from the opaque body of a balloon. It was a red body, with greatest luminosity in its nucleus. The Editor of *Cosmos* writes that, according to other correspondents, this object had been seen, at 11 P.M., July 19th and 26th, at Chatou. Argenteuil and Chatou are 4 or 5 miles apart, and both are about 5 miles from Paris. All three of these dates were Sundays, and even though nothing like a balloon had been seen through glasses, one naturally supposes that somebody near Paris had been amusing himself sending up fire-balloons, Sunday evenings. The one great resistance to all that is known as progress is what one "naturally supposes."

In the *English Mechanic*, 81-220, Arthur Mee writes that several persons, in the neighborhood of Cardiff, had, upon the night of March 29, 1905, seen in the sky, "an appearance like a vertical beam of light, which was not due (they say) to a searchlight, or any such cause." There were other observations, and they remind us of the observations by Noble and Bradgate, Aug. 28-29, 1883: then upon an object that cast a light like a searchlight; this time an association between a light like a searchlight, and a luminosity of definite form. In the *Cambrian Natural Observer*, 1905-32, are several accounts of a more definite-looking appearance that was seen, this night, in the sky of Wales —"like a long cluster of stars, obscured by a thin film or mist." It was seen at the time of the visibility of Venus, then an "evening star"—about 10 P.M. It grew brighter, and for about half an hour looked like an incandescent light. It was a conspicuous and definite object, according to another description—"like an iron bar, heated to an orange-colored glow, and suspended vertically."

Three nights later, something appeared in the sky of Cherbourg, France— *L'Astre Cherbourg*—the thing that appeared, night after night, in the sky of the city of Cherbourg, at a time when the planet Venus was nearest (inferior conjunction April 26, 1905).

Flammarion, in the *Bull. Soc. Astro. de France*, 19-243, says that this object was the planet Venus. He therefore denies that it had moved in various directions, saying that the supposed observations to this effect were illusions. In *L'Illustration*, April 22, 1905, he tells the story in his own way, and says some things that we are not disposed to agree with, but also he says that the ignorance of some persons in *inénarrable*. In *Cosmos*, n.s., 42-420, months after the occurrence, it is said that many correspondents had written to inquire as to *L'Astre Cherbourg*. The Editor gives his opinion that the object was either Jupiter or Venus. Throughout our Venus-visitor expression, the most important point is appearance in a local sky. That unifies this expression with other expressions, all of them converging into our general extra-geographic acceptances. The Editor of Cosmos says that this object, which was reported from Cherbourg, was reported from other towns as well. He probably means to say that it was seen simultaneously in different towns. For all guardians of this earth's isolation, this is a convenient thing to say: the conclusion then is that the planet Venus, exceptionally bright, was attracting unusual attention generally, and that there was nothing in the especial sky of Cherbourg. But we have learned that standardizing disguisements often obscure our data in later accounts, and we have formed the habit of going to contemporaneous sources. We shall find that the newspapers of the time reported a luminous object that appeared, night after night, only over the city of Cherbourg, as the name by which it was known indicates. It was a reddish object. The Editor of*Cosmos* explains that atmospheric conditions could give this coloration to Venus. I suppose this could be so occasionally: not night after night, I should say. We shall find that this object, or a similar object, was reported from other places, but not simultaneously with its appearance over Cherbourg.

In the *Journal des Debats*, the first news is in the issue of April 4, 1905. It is said that a luminous body was appearing, every evening, between 8 and to o'clock, over the city of Cherbourg.

These were about the hours of the visibility of Venus. In this period, Venus set at 9:30 P.M., and Jupiter at 8 P.M. It is enough to make any conventionalist feel most reasonable, though he'd feel that way anyway, in thinking that of course then this object was Venus. In my own earlier speculations upon this subject, this one datum stood out so that had it not been for other data, I'd have abandoned the subject. But then I read, of other occurrences: time after time has something been seen in a local sky of this earth, sometimes so definitely seen to move, not like Venus, but in various directions, that one has to think that it was not Venus, though appearing at the time of visibility of Venus. Between these appearances and visibility of Venus there does seem to be relation.

In the Journal, it is said that *L'Astre Cherbourg* had an apparent diameter of 15 centimeters, and a less definite margin of 75 centimeters—seemed to be about a yard wide—meaningless of course. In the *Bull. Soc. Astro. de France*, it is said that, according to reports, its form was oval. In the journal *des Debats*, we are told that at first the thing was supposed to be a captive balloon but that this idea was given up because it appeared and disappeared.

Journal des Debats, April 12:

That every evening the luminous object was continuing to appear above Cherbourg; that many explanations had been thought of: by some persons that it was the planet Jupiter, and by others that it was a comet but that no one knew what it was. The comet-explanation is of course ruled out. The writer in the journal expresses regret that neither the Meteorological Bureau nor the Observatory of Paris had sent anybody to investigate, but says that the *préfet maritime* of Cherbourg had commissioned a naval officer to investigate. In *Le Temps*, of the 12th, is published an interview with Flammarion, who complains some more against general *inénarrable-ness*, and says that of course the object was Venus. The writer in *Le Temps* says that soon would the matter be settled, because the commander of a warship had undertaken to decide what the luminous body was. *Le Figaro*, April 13:

The report of Commander de Kerillis, of the *Chasseloup-Laubut*—that the position of *L'Astre Cherbourg* was not the position of Venus, and that the disc did not look like the crescentic disc of Venus, but that the observations had been made from a vessel, under unfavorable conditions, and that the commander and his colleagues did not offer a final opinion.

I think that there was *inénarrable-ness* all around. Given visibility, I can't think what the unfavorable conditions could have been. Given, however, observations upon something that all the astronomers in the world would say could not be, one does think of the dislike of a naval officer, who, though he probably knew right ascension from declination, was himself no astronomer, to commit himself. In *Le Temps*, and other newspapers published in Paris, it is said that, according to the naval officers, the object might have been a comet, but that they would not positively commit themselves to this opinion, either.

I think that somebody should be brave; so, though not positively, of course, I incline, myself, to relate these appearances over Cherbourg with the observations in Wales, upon March 29th; also I suggest that there is another report that may relate. In *Le Temps*, April 12, it is said that, at midnight, April 9-10, a luminous body, like *L'Astre Cherbourg*, was seen in the sky of Tunis. Though it was visible several minutes, it is said that this object was probably a meteor.

Every night, from the first to the eleventh of April, a luminous body appeared in the sky of Cherbourg. Then it was seen no longer. It may have been seen sailing away, upon its final departure from the sky of Cherbourg. In *Le Figaro*, April 15, it is said that, upon the night of the eleventh of April, the guards of La Blanche Lighthouse had seen something like a lighted balloon in the sky. Supposing it was a balloon, they had started to signal to it, but it had disappeared. It is said that the lighthouse had been out of communication with the mainland, and that the guards had not heard of *L'Astre Cherbourg*.

In the London *Times*, Nov. 23, 1905, a correspondent writes that, at East Liss, Hants, which is about 40 miles from Reading, he and his gamekeeper had, about 3:30 P.M., Nov. 17th, heard a loud, distant rumbling. According to this hearer, the rumbling seemed to be a composition of triplets of sounds. We shall accept that three sounds were heard, but we have no other assertion that each sound was itself so sub-serialized. This correspondent's gamekeeper said that he had heard similar sounds at 11:30 A.M., and at 1:30 P.M. It is said that the sounds were not like gunfire, and that the direction from which they seemed to come, and the time in the afternoon, precluded the explanation of artillery-practice at Aldershot or Portsmouth. Aldershot is about 15 miles from East Liss, and Portsmouth about 20.

Times, November 24—that the "quake" had been distinctly felt in Reading, about 3:30 P.M., November 17th. *Times*, November 25—heard at Reading, at 11:30, 1:30, and 3:30 o'clock, November 17th.

Reading Standard, November 25:

That consternation had been caused in Reading, upon the 17th, by sounds and vibrations of the earth, about 11:30 A.M., 1:30 P.M., and 3:30 P.M. It is said that nothing had been seen, but that the sounds closely resembled those that had been heard during the meteoric shower of 1866.

Mr. H. G. Fordham appears again. In the *Times*, December 1, he writes that the phenomena pointed clearly to an explosion in the sky, and not to an earthquake of subterranean origin. "The noise and shock experienced are no doubt attributable to the explosion (or to more than one explosion) of a meteorite, or bolide, high up in the atmosphere, and setting up a wave (or waves) of sound and aërial shock. It is probable, indeed, that a good many phenomena having this source are wrongly ascribed to slight and local earth-shock."

Mr. Fordham wrote this, but he wrote no more, and I think that somewhere else something else was written, and that, in the year 1905, it had to be obeyed; and that it may be interpreted in these words—"Thou shalt not." Mr. Fordham did not inquire into the reasonableness of thinking that, only by coincidence, meteors so successively exploded, in a period of four hours, in one local sky of this earth, and nowhere else; and into the inference, then, as to whether this earth is stationary or not.

We have data of a succession occupying far more than four hours.

In the *Times*, Mrs. Lane, of Petersfield, 20 miles from Portsmouth, writes that, at 11:30 A.M., and at 3:30 P.M., several days before the 17th, she had heard the detonations, then hearing them again, upon the 17th. Mrs. Lane thinks that there must have been artillery-practice at Portsmouth. It seems clear that there was no cannonading anywhere in England, at this time. It seems clear that there was signaling from some other world.

In the *English Mechanic*, 82-433, Joseph Clark writes that, a few minutes past 3 P.M., upon the 18th a triplet of detonations was heard at Somerset—"as loud as thunder, but not exactly like thunder."

Reading Observer, November 25—that, according to a correspondent, the sounds had been heard again, at Whitechurch (20 miles from Reading) upon the 21st, at 1:35 P.M., and 3:08 P.M. The sounds had been attributed to artillery-practice at Aldershot, but the correspondent had written to the artillery commandant, at Bulford Camp, and had received word that there had been no heavy firing at the times of his inquiry. The Editor of the *Observer* says that he, too, had written to the commandant, and had received the same answer.

I have searched widely. I have found record of nobody's supposition that he had traced these detonations to origin upon this earth.

33

In Coconino County, Arizona, is an extraordinary formation. It is known as Coon Butte and as Crater Mountain. Once upon a time, something gouged this part of Arizona. The cavity in the ground is about 3,800 feet in diameter, and it is approximately 600 feet deeps from the rim of the ramparts to the floor of the interior. Out from this cavity had been hurled blocks of limestone, some of them a mile or so away, some of these masses weighing probably 5,000 tons each. And in the formation, and around it, have been found either extraordinary numbers of meteorites, or fragments of one super-meteorite. Barringer, in his report to the Academy of Natural Science of Philadelphia (*Proceedings, A. N. S. P.*, December, 1905) says that, of the traffickers in this meteoritic material, he knew of two men who had shipped away fifteen tons of it. But Barringer's minimum estimate of a body large enough so to gouge the ground is ten million tons.

It was supposed that a main mass of meteoritic material was buried under the floor of the formation, but this floor was drilled, and nothing was found to support this supposition. One drill went down 1,020 feet, going through too feet of red sandstone, which seems to be the natural, undisturbed sub-structure. The datum that opposes most strongly the idea that this pit was gouged by one super-meteorite is that in it and around it at least three kinds of meteorites have been found: they are irons, masses of iron-shale, and shale-balls that are so rounded and individualized that they cannot be thought of as fragments of a greater body, and cannot be very well thought of as great drops of molten matter cast from a main, incandescent mass, inasmuch as there is not a trace of igneous rock such as would mark such contact.

There are data for thinking that these three kinds of objects fell at different times, presumably from origin of fixed position relatively to this point in Arizona. Within the formation, shales were found, buried at various distances, as if they had fallen at different times, for instance seven of them in a vertical line, the deepest-buried 27 feet down; also shales outside the formation were found buried. But, quite as if they had fallen more recently, the hundreds of irons were found upon the surface of the ground, or partly covered, or wholly covered, but only with superficial soil.

There is no knowing when this great gouge occurred, but cedars upon the rim are said to be about 700 years old.

In terms of our general expression upon differences of potential, and of electric relations between nearby worlds, I think of a blast between this earth and a land somewhere else, and of something that was more than a cyclone that gouged this pit.

Other meteorites have been found in Arizona: the 85-pound iron that was found at Weaver, near Wickenburg, 130 miles from Crater Mountain, in 1898, and the 960-pound mass, now in the National Museum, said to have been found at Peach Springs, 140 miles from Crater Mountain. These two irons indicate nothing in particular; but, if we accept that somewhere else in Arizona there is another deposit of meteorites, also extraordinarily abundant, such abundance gives something of commonness of nature if not of commonness of origin to two deposits. There are several large irons known as the Tucson meteorites, one weighing 632 pounds and another 1,514 pounds, now in museums. They came from a place known as Iron Valley, in the Santa Rita Mountains, about 30 miles south of Tucson, and about 200 miles from Crater Mountain. Iron Valley was so named because of the great number of meteorites found in it. According to the people of Tucson, this fall occurred about the year 1660. See *Amer. Jour. Sci.*, 2-13-290.

Upon June 24, 1905, Barringer found, upon the plain, about a mile and a half northwest of Crater Mountain, a meteorite of a fourth kind. It was a meteoritic stone, "as different from all the other specimens as one specimen could be from another." Barringer thinks that it fell, about the 15th of January, 1904. Upon a night in the middle of January, 1904, two of his employees were awakened by a loud hissing sound, and saw a meteor falling north of the formation. At the same time, two Arizona physicians, north of the formation, saw the meteor falling south of them. For analysis and description of this object, see *Amer. Jour. Sci.*, 4-21-353. Barringer, who believes that once upon a time one super-meteorite, of which only a very small part has ever been found, gouged this hole in the ground, writes—"That a small stony meteorite should have fallen on almost exactly the same spot on this earth's surface as the great Canon Diablo iron meteorite fell many centuries ago, is certainly a most remarkable coincidence. I have stated the facts as accurately as possible, and I have no opinion to offer, as to whether or not these involve anything more than a coincidence."

Other phenomena in Arizona:

Upon Feb. 24, 1897, a great explosion was heard over the town of Tombstone. It is said that a fragment of a meteor fell at St. David (*Monthly Weather Review*, 1897-56). Yarnell, Arizona, Sept. 12, 1898—"a loud, deep, thundering noise" that was heard between noon and 1 P.M. "The noise proceeded from the Granite Range, this side of Prescott. From all accounts, a large meteor struck the earth at this time" (*U. S. Weather Bureau Rept., Ariz. Section*, September, 1898).

Upon July 19, 1912, at Holbrook, Arizona, about 50 miles from Crater Mountain, occurred a loud detonation and one of the most remarkable falls of stones recorded. See *Amer. Jour. Sci.*, 4-34-437. Some of the stones are very small. About 14,000 were collected. Only twice, since the year 1800, have stones in greater numbers fallen from the sky to this earth, according to conventional records.

About a month later (August 18) there was another concussion at Holbrook. This was said to be an earthquake (*Bull. Seis. Soc. Amer.*, 1-209).

34

THE climacteric opposition of Mars, of 1909—the last in our records—the next will be in 1924—

Aug. 8, 1909—see *Quar. Jour. Met. Soc.*, n.s., 35-299—flashes in a clear sky that were seen in Epsom, Surrey, and other places in the southeast of England. They could not be attributed to lightning in England. The writer in the Journal finds that there was a storm in France, more than one hundred miles away. For an account of these flashes, tabulated at Epsom—"night fine and starlight"—see *Symons' Met. Mag.*, 44-148. During each period of five minutes, from 10 to 11:15 P.m., the number of flashes-16-14-20-31-15-26-12-20-30-18-27-22-14-12-10-21-8-5-3-1-0-1-0. With such a time-basis, I can see no possibility of detecting anything of a code-like significance. I do see development. There were similar observations at times in the favorable oppositions of Mars of 1875 and 1877. In 1892, such flashes were noted more particularly. Now we have them noted and tabulated, but upon a basis that could be of interest only to meteorologists. If they shall be seen in 1924, we may have observation, tabulation, and some marvelously different translations of them. After that there will be some intolerably similar translations, suspiciously delayed in publication.

Sept. 23, 1909—opposition of Mars.

Throughout our data, we have noticed successions of appearances in local skies of this earth, that indicate that this earth is stationary, but that also relate to nearest approaches of Mars. Upon the night of Dec. 16-17, 1896, concussion after concussion was felt at Worcester, England; a great "meteor" was seen at the time of the greatest concussion. Mars was seven days past opposition. We thought it likely enough that explosion after explosion had occurred over Worcester, and that something in the sky had been seen only at the time of the greatest, or the nearest, explosion. We did not think well of the conventional explanation that only by coincidence had a great meteor exploded over a region where a series of earthquakes was occurring, and exactly at the moment of the greatest of these shocks.

In November, 1911, Mars was completing its cycle of changing proximities of a duration of fifteen years, and was duplicating the relationship of the year 1896. About to o'clock, night of Nov. 16, 1911, a concussion that is conventionally said to have been an earthquake occurred in Germany and Switzerland. But plainly there was an explosion in the sky. In the *Bulletin of the Seismological Society of America,* 3-189, Count Montessus de Ballore writes that he had examined 112 reports upon flashes and other luminous appearances in the sky that had preceded the "earthquake" by a few seconds. He concludes that a great meteor had only happened to explode over a region where, a few seconds later, there was going to be an earthquake. "It therefore seems highly probable that the earthquake coincided with a fall of meteors or of shooting stars."

The duplication of the circumstances of December, 1896, continues. If of course this concussion in Germany and Switzerland was the effect of something that exploded in the sky—of what were the concussions that were felt later, the effects? De Ballore does not mention anything that occurred later.

But, a few minutes past midnight, and then again, at 3 o'clock, morning of the 17th, there were other, but slighter, shocks. Only at the time of the greatest shock was something seen in the sky.*Nature*, 88-117—that this succession of phenomena did occur. We relate the phenomena to the planet Mars, but also we ask—how, if most reasonably, all three of these shocks were concussions from explosions in the sky, if of course one of them was, meteors could ever so hound one small region upon a moving earth, or projectiles be fired with such specialization and preciseness? Nov. 17th, 1911, was seven days before the opposition of Mars. Though the opposition occurred upon the 24th of November, Mars was at minimum distance upon the 17th.

No matter how difficult of acceptance our own notions may be, they are opposed by this barbarism, or puerility, or pill that can't be digested:

Seven days from the opposition of Mars, in 1896, a great meteor exploded over a region where there had been a succession of earthquakes—by coincidence;

Seven days from the next similar opposition of Mars, a great meteor exploded over a region where there was going to be a succession of earthquakes—by coincidence.

The Advantagerians of the moon—that is the cult of lunar cornmunicationists, who try to take advantage of such celestial events as oppositions and eclipses, thinking that astronomers, or night watchmen, or policemen of this earth might at such times look up at the sky

A great luminous object, or a meteor, that was seen at the time of the eclipse of June 28, 1908—"as if to make the date of the eclipse more memorable," says W. F. Denning (*Observatory*, 31-288).

Not long before the opposition of Mars, in 1909, the bright spot west of Picard was seen twice: March 26 and May 23 (*Jour. B. A. A.*, 19-376).

Nov. 16, 1910—an eclipse of the moon, and a "meteor" that appeared, almost at the moment of totality (*Eng. Mec.*, 92-430). It is reported, in *Nature*, 85-118, as seen by Madame de Robeck, at Naas, Ireland, "from an apparent radiant, just below the eclipsed moon." The thing may have come from the moon. Seemingly with the same origin, it was seen far away in France. In*La Nature*, Nov. 26, 1910, it is said that, at Besançon, France, during the eclipse, was seen a meteor like a superb rocket, "qui serait partie de la lune." There may have been something occurring upon the moon at the time. In the *Jour. B. A. A.*, 21-100, it is said that Mrs. Albright had seen a luminous point upon the moon throughout the eclipse.

Our expression is that there is an association between reported objects, like extra-mundane visitors, and nearest approaches by the planet Venus to this earth. Perhaps unfortunately this is our expression, because it makes for more restriction than we intend. The objects, or the voyagers, have often been seen during the few hours of the visibility of Venus, when the planet is nearest. "Then such an object is Venus," say the astronomers. If anybody wonders why, if these seeming navigators can come close to this earth—as they do approach, if they appear only in a local sky—they do not then come all the way to this earth, let him ask a sea captain why said captain never purposely descends to the bottom of the ocean, though traveling often not far away. However, I conceive of a great variety of extra-mundanians, and I am now collecting data for a future expression—that some kinds of beings from outer space can adapt to our conditions, which may be like the bottom of a sea, and have been seen, but have been supposed to be psychic phenomena.

Upon Oct. 31, 1908, the planet Venus was four months past inferior conjunction, and so had moved far from nearest approach, but there are vague stories of strange objects that had been seen in the skies of this earth— localized in New England—back to the time of nearest approach. In the *New York Sun*, Nov. 1, 1908, is published a dispatch, from Boston, dated Oct. 31. It is said that, near Bridgewater, at four o'clock in the morning of October 31, two men had seen a spectacle in the sky. The men were not astronomers. They were undertakers. There may be a disposition to think that these observers were not in their own field of greatest expertness, and to think that we are not very exacting as to the sources of our data. But we have to depend upon undertakers, for instance: early in our investigations, we learned that the prestige of astronomers has been built upon their high moral character, all of them most excellently going to bed soon after sunset, so as to get up early and write all day upon astronomical subjects. But the exemplary in one respect may not lead to much advancement in some other respect. Our undertakers saw, in the sky, something like a searchlight. It played down upon this earth, as if directed by an investigator, and then it flashed upward. "All of the balloons in which ascensions are made, in this State, were accounted for today, and a search through southeastern Massachusetts failed to reveal any further trace of the supposed airship." It is said that "mysterious bright lights," believed to have come from a balloon, had been reported from many places in New England. The week before, persons at Ware had said that they had seen an illuminated balloon passing over the town, early in the morning. During the summer such reports had come from Bristol, Conn., and later from Pittsfield, Mass., and from White River Junction, Vt. "In all these cases, however, no balloon could be found, all the known airships being accounted for." In the *New York Sun*, Dec. 13, 1909, it is said that, during the autumn of 1908, reports had come from different places in Connecticut, upon a mysterious light that moved rapidly in the sky.

Venus moved on, traveling around the sun, which was revolving around this earth, or traveling any way to suit anybody. In December, 1909, the planet was again approaching this earth. So close was Venus to this earth that, upon the 15th of December, 1909, crowds stood, at noon, in the streets of Rome, watching it, or her (*New York Sun*, December 16). At 3 o'clock, afternoon of December 24th crowds stood in the streets of New York, watching Venus (*New York Tribune*, December 25). One supposes that upon these occasions Venus may have been within several thousand miles of this earth. At any rate I have never heard of one fairly good reason for supposing otherwise. If again something appeared in local skies of this earth, or in the skies of New England, and sometimes during the few hours of the visibility of Venus, the object was or was not Venus, all according to the details of various descriptions, and the credibility of the details. The searchlight, for instance; more than one light; directions and motions. Venus, at the time, was for several hours after sunset, slowly descending in the southwest: primary maximum brilliance Jan. 8th, 1910; inferior conjunction February 12th.

There is an amusing befuddlement to clear away first. Upon the night of Sept. 8, 1909, a luminous object had been seen sailing over New England, and sounds from it, like sounds from a motor, had been heard. Then Mr. Wallace Tillinghast, of Worcester, Mass., announced that this light had been a lamp in his "secret aeroplane," and that upon this night he had traveled, in said "secret aeroplane," from Boston to New York, and back to Boston. At this time the longest recorded flight, in an aeroplane, was Farman's, of 111 miles, from Rheims, August, 1909; and, in the United States, according to records, it was not until May 29, 1910, that Curtiss flew from Albany to New York City, making one stop in the 150 miles, however. So this unrecorded flight made some stir in the newspapers. Mr. Tillinghast meant his story humorously of course. I mention it because, if anybody should look the matter up, he will find the yarn involved in the newspaper accounts. If nothing else had been seen, Mr. Tillinghast might still tell his story, and explain why he never did anything with his astonishing "secret aeroplane"; but something else was seen, and upon one of the nights in which it appeared, Tillinghast was known to be in his home.

According to the *New York Tribune*, Dec. 21, 1909, Immigration Inspector Hoe, of Boston, had reported having seen, at one o'clock in the morning of December 20, "a bright light passing over the harbor" and had concluded that he had seen an airship of some kind.

New York Tribune, December 23—that a "mysterious airship" had appeared over the town of Worcester, Mass., "sweeping the heavens with a searchlight of tremendous power." It had come from the southeast, and traveled northwest, then hovering over the city, disappearing in the direction of Marlboro. Two hours later, it returned. "Thousands thronged the streets, watching the mysterious visitor." Again it hovered, then moving away, heading first to the south and then to the east.

The next night, something was seen, at 6 o'clock, at Boston. "The searchlights shot across the sky line." "As it flew away to the north, queries began to pour into the newspaper offices and the police stations, regarding the remarkable visitation." It is said that an hour and a half later, an object that was supposed to be an airship with a powerful searchlight, appeared in the sky, at Willimantic, Conn., "hovering" over the town about 15 minutes. In the *New York Sun*, December 24, are more details. It is said that, at Willimantic, had been seen a large searchlight, approaching from the east, and that then dark outlines of something behind the searchlight had been seen. Also, in the Sun, it is said that whatever it may have been that was seen at Boston, it was a dark object, with several red lights and a searchlight, approaching Boston from the west, hovering for 10 minutes, and then moving away westward. From Lynn, Mass., it was described as "a long black object," moving in the direction of Salem, and then returning, "at a high speed." It is said that the object had been seen at Marlboro, Mass., nine times since December 14.

New York Tribune, Jan. 1, 1910—dispatch from Huntington, West Virginia, Dec. 31, 1909—"Three huge lights of almost uniform dimensions appeared in the early morning sky, in this neighborhood, today. Joseph Green, a farmer, declared that they were meteors, which fell on his farm. An extensive search of his land by others who saw the lights was fruitless, and many persons believe that an airship had sped over the country."

In the Tribune, Jan. 13, 1910, it is said that, at 9 o'clock, morning of January 12, an airship had been seen at Chattanooga, Tenn. "Thousands saw the craft, and heard the 'chug' of its engine." Later the object was reported from Huntsville, Alabama.*New York Tribune*, January 15—dispatch from Chattanooga, January 14—"For the third successive day, a mysterious white aircraft passed over Chattanooga, about noon today. It came from the north, and was traveling southeast, disappearing over Missionary Ridge. On Wednesday, it came south, and on Thursday, it returned north."

In the middle of December, 1909, someone had won a prize for sailing in a dirigible from St. Cyr to the Eiffel Tower and back.

St. Cyr is several miles from Paris.

Huntsville, Ala., and Chattanooga, Tenn., are 75 miles apart.

An association between the planet Venus and "mysterious visitors" either illumines or haunts our data. In the *New York Tribune*, Jan. 29, 1910, it is said that a luminous object, thought to be Winnecke's comet, had been seen,

January 28, near Venus; reported from the Manila Observatory.

I have another datum that perhaps belongs to this series of events. Every night, from the 14th to the 23rd of December, 1909, if we accept the account from Marlboro, a luminous object was seen traveling, or exploring, in the sky of New England. Certainly enough it was no "secret airship" of this earth, unless its navigator went to extremes with the notion that the best way to kept a secret is to announce it with red lights and a searchlight. However, our acceptance depends upon general data as to the development of terrestrial aeronautics. But upon the night of December 24th, the object was not seen in New England, and it may have been traveling or exploring somewhere else. Night of the 24th—Venus in the southwest in the early hours of the evening. In the *English Mechanic*, 104-71, a correspondent, who signs himself "Rigel," writes that, upon December 24, at 8:30 o'clock in the evening, he saw a luminous object appear above the northeastern horizon and slowly move southward, until 8.50 o'clock, then turning around, retracing, and disappearing whence it came, at two minutes past nine. The correspondent is James Fergusen, Rossbrien, Limerick, Ireland. He writes frequently upon astronomical and meteorological subjects, and is still contributing to the somewhat enlightened columns of the *English Mechanic*.

Nov. 19, 1912—explosive sounds reported from Sunninghill, Berkshire. No earthquake was recorded at the Kew Observatory, and, in the opinion of W. F. Denning (*Nature*, 9-363, 417) the explosion was in the sky. It was a terrific explosion, according to the *Westminster Gazette* (November 19). There was either one great explosion that rumbled and echoed for five minutes, or there were repeated detonations, resembling cannonading—"like a tremendous discharge of big guns" according to reports from Abingdon, Lewes, and Epsom. Sunninghill is about ten miles from Reading, and Abingdon is near Reading, but the sound was heard in London, and down by the English Channel, and even in the island of Alderney. In the *Gazette*, November 28, Sir George Fordham (H. G. Fordham) writes that, in his opinion, it was an explosion in the sky. He says—"The phenomena of airshock never have, I believe, been very fully investigated." His admissions and his omissions remain the same as they have been since occurrences of the year 1889. He does not mention that, according to Philip T. Kenway, of Hambledon, near Godalming, about thirty miles southeast of Reading, the sounds were heard again the next day, from 1:45 to 2 P.M. Mr. Kenway thinks that there had been big-gun firing at Portsmouth (*West. Gaz.*, November 21). In the London *Standard*, a correspondent, writing from Dorking, say that the phenomena of the 19th were like concussions from cannonading—"at regular intervals"—"at quick intervals, lasting some seconds each time, for five minutes, by the clock."

It develops that Reading was the center over which the detonations occurred. In the *Westminster Gazette*, November 30, it is said that the shocks had been felt in Reading, upon the 19th, 20th, and 21st. Only from Reading have I record of phenomena upon the 21st. Mr. H. L. Hawkins, Lecturer in Geology, of the Reading University, writes that according to his investigations there had been no gun-firing in England, to which the detonations could be attributed. He says that Fordham's explanation was in accord with his own investigations, or that the detonations had occurred in the sky. He writes that, inasmuch as the detonations had occurred upon three successive days, a shower of meteors, of long duration, would have to be supposed. How he ever visualized that unerring shower, striking one point over this earth's surface, and nowhere else, day after day, if this earth be a rotating and revolving body, I cannot see. If he should say that by coincidence this repetition could occur, then by what coincidence of coincidences could the same repetitions have occurred in this same local sky, centering around Reading, seven years before? The indications are that this earth is stationary, no matter how unreasonable that may sound.

In the *Westminster Gazette*, December 9, W. F. Denning writes that without doubt the phenomena were "meteoric explosions." But he alludes to the "airquake and strange noises" that were heard upon the 19th. He does not mention the detonations that were heard upon the following days. Not one of these writers mentions the sounds that were heard in Reading, in November, 1905.

London *Standard*, Nov. 23, 1912—that, according to Lieut. Col. Trewman, of Reading, the sounds had been heard at Reading, at 9 A.M., upon the 19th; 1:45 P.M., the 10th; 3:30 P.M., the 21st.

35

"Unknown Aircraft Over Dover."

According to the Dover correspondent to the London *Times* (Jan. 6, 1913) something had been seen, over Dover, heading from the sea.

In the London *Standard*, Jan. 24, 1913, it is said that, upon the morning of January 4, an unknown airship had been seen, over Dover, and that, about the same time, the lights of an airship had been seen over the Bristol Channel. These places are several hundred miles apart.

London *Times*, January 21—report by Capt. Lindsay, Chief Constable of Glamorganshire: that, about five o'clock, in the afternoon of January 17, he saw an object in the sky of Cardiff, Wales. He says that he called the attention of a bystander, who agreed with him that it was a large object. "It was much larger than the Willows airship, and left in its trail a dense smoke. It disappeared quickly."

The next day, according to the *Times*, there were other reports: people in Cardiff saw something that was lighted or that carried lights, moving rapidly in the sky. In the *Times*, of the 28th, it is said that an airship that carried a brilliant light had been seen in Liverpool. "It is stated at the Liverpool Aviation School that none of the airmen had been out on Saturday night." Dispatches from town after town—a traveling thing in the sky, carrying a light, and also a searchlight that swept the ground. It is said that a vessel, of which the outlines had been clearly seen, had appeared in the sky of Cardiff, Newport, Neath, and other places in Wales. In the *Standard*, January 31, is published a list of cities where the object had been seen. Here a writer tries to conclude that some foreign airship had made half a dozen visits to England and Wales, or had come once, remaining three weeks; but he gives up the attempt, thinking that nothing could have reached England and have sailed away half a dozen times without being seen to cross the coast; thinking that the idea of anything having made one journey, and remaining three weeks in the air deserved no consideration.

If the unknown object did carry something like a searchlight, an idea of its powers is given in an account in the *Cardiff Evening Express*, Jan. 25, 1913 —"Last evening brilliant lights were seen, sweeping skyward, and now, this evening, the lights grow bolder. Streets and houses in the locality of Totterdown were suddenly illuminated by a brilliant, piercing light, which, sweeping upward, gave many spectators a fine view of the hills beyond." In the *Express*, February 6, is a report upon this light like a searchlight, and the object that flashed it, by the police of Dulais Valley. Also there is an account, by a police sergeant, of a luminous thing that was for a while stationary in the sky, and then moved away. Still does the conventional explanation, or suggestion, survive. It is said that members of the staff of the Evening Express had gone to the roof of the newspaper building, but had seen only the planet Venus, which was brilliant at this time.

Then writes a correspondent, to the *Express*, that the object could not have been Venus, because he had seen it traveling at a rate of 20 or 30 miles an hour, and had heard sounds from it. Someone else writes that not possibly could the thing be Venus: he had seen it as "a bright red light, going very fast." Still someone else says that he had seen the seeming vessel upon the 5th of February, and that it had suddenly disappeared.

There is a hiatus. Between the 5th and the 21st of February, nothing like an airship was seen in the sky of England and Wales. If we can find that somewhere else something similar was seen in the sky, in this period, one supposes that it was the same object, exploring or maneuvering somewhere else. It seems however that there were several of these objects, because of simultaneous observations at places far apart. If we can find that, during the absence from England and Wales, similar objects were seen somewhere else, a great deal of what we try to think upon the subject will depend upon how far from Great Britain they were seen. It seems incredible that the planet Venus should deceive thousands of Britons, up to the 5th of February, and stop her deceptions abruptly upon that date, and then abruptly resume deceptions upon the 21st, in places at a distance apart. These circumstances oppose the idea of collective hallucinations, by which some writers in the newspapers tried to explain. If they were hallucinations, the hallucinations renewed collectively, upon the 21st, in towns one hundred miles apart. One extraordinary association is that all appearances, except the first, were in the hours of visibility of Venus, then an "evening star."

Upon the night of the 21st, a luminous object was reported from towns in Yorkshire and from towns in Warwickshire, two regions about one hundred miles apart; about 10 P.M. All former attempts to explain had been abandoned, and the general supposition was that German airships were maneuvering over England. But not a thing had been seen to cross the coast of England, though guards were patroling the coasts, especially commissioned to watch for foreign airships. Sailors in the North Sea, and people in Holland and Belgium had seen nothing that could be thought a German airship sailing to or from England. A writer in*Flight* takes up as especially mysterious the appearance far inland, in Warwickshire. Then came reports from Portsmouth, Ipswich, Hornsea, and Hull, but, one notes, no more, at this time, from Wales. Also in Ipswich, which is more than a hundred miles from the towns in Warwickshire, and more than a hundred miles from the Yorkshire towns, a luminous object was seen upon the night of the 21st. *Ipswich Evening Star*, February 25— something that carried a searchlight that had been seen upon the nights of the 21st and 24th, moving in various directions, and then "dashing off at lightning speed"—that, at Hunstanton, had been seen three bright lights traveling from the eastern sky, remaining in sight 30 minutes, stationary, or hovering over the town, and then disappearing in the northwest. *Portsmouth Evening News*, February 25—that soon after 8 P.M., evening of the 24th, had been seen a very bright light, appearing and disappearing, remaining over Portsmouth about one hour, and then moving away. Portsmouth and Ipswich are about 120 miles apart. In the London newspapers, it is said that, upon the evening of the 25th, crowds stood in the streets of Hull, watching something in the sky, "the lights of which were easily distinguishable." Hull is about 190 miles northeast of Portsmouth. *Hull Daily Mail*, February 26—that a crowd had watched a light high in the air. It is said that the light had been stationary for almost half an hour and had then shot away northward. In the Times, February 28, are published reports upon "the clear outlines of an airship, which was carrying a dazzling searchlight," from Portland, Burcleaves, St. Alban's Head, Papplewich, and the Orkneys. The last account, after a long interval, that I know of, is another report from Capt. Lindsay: that, about 9 o'clock, evening of April 8th, he and many other persons had seen, over Cardiff, something that carried a brilliant light and traveled at a rate of sixty or seventy miles an hour.

Upon April 24, 1913, the planet Venus was at inferior conjunction.

In the *Times*, February 28, it is said that a fire-balloon had been found in Yorkshire, and it is suggested that someone had been sending up fire-balloons.

In the *Bull. Soc. Astro. de France*, 1913-178, it is said that the people of England were as credulous as the people of Cherbourg, and had permitted themselves to be deceived by the planet Venus.

If German airships were maneuvering over England, without being seen either approaching or departing, appearing sometimes far inland in England without being seen to cross the well-guarded coasts, it was secret maneuvering, inasmuch as the accusation was denied in Germany (*Times*, February 26 and 27). It was then one of the most brilliantly proclaimed of secrets, or it was concealment under one of the most powerful searchlights ever seen. Possibly an airship from Germany could appear over such a city as Hull, upon the east coast of England, without being seen to arrive or to depart, but so far from Germany is Portsmouth, for instance, that one does feel that something else will have to be thought of. The appearances over Liverpool and over towns in Wales might be attributed to German airships by someone who has not seen a map since he left school. There were more observations upon sudden appearances and disappearances than I have recorded: stationariness often occurred.

The objects were absent from the sky of Great Britain, from February 5 to February 21.

According to data published by Prof. Chant, in the *Journal of the Royal Astronomical Society of Canada*, 7-148, the most extraordinary procession in our records was seen, in the sky of Canada, upon the night of Feb. 9, 1913. Either groups of meteors, in one straight line, passed over the city of Toronto, or there was a procession of unknown objects, carrying lights. According to Prof. Chant, the spectacle was seen from the Saskatchewan to Bermuda, but if this long route was traversed, data do not so indicate. The supposed route was diagonally across New York State, from Buffalo, to a point near New York City, but from New York State are recorded no observations other than might have been upon ordinary meteors, this night. A succession of luminous objects passed over Toronto, night of Feb. 9, 1913, occupying from three to five minutes in passing, according to different estimates. If one will think that they were meteors, at least one will have to think that no such meteors had ever been seen before. In the *Journal*, 7-405, W. F. Denning writes that, though he had been watching the heavens since the year 1865, he had never seen anything like this. In most of the observations, the procession is described as a whole—"like an express train, lighted at night"—"the lights were at different points, one in front, and a rear light, then a succession of lights in the tail." Almost all of the observations relate to the sky of Toronto and not far from Toronto. It is questionable that the same spectacle was seen in Bermuda, this night. The supposed long flight from the Saskatchewan to Bermuda might indicate something of a meteoric nature, but the meteor-explanation must take into consideration that these objects were so close to this earth that sounds from them were heard, and that, without succumbing to gravitation, they followed the curvature of this earth at a relatively low velocity that cannot compare with the velocity of ordinary meteors.

If now be accepted that again, the next day, objects were seen in the sky of Toronto, but objects unlighted, in the daytime—I suppose that to some minds will come the thought that this is extraordinary, and that almost immediately the whole subject will

t then be forgotten. Prof. Chant says that, according to the *Toronto Daily Star*, unknown objects, but dark objects this time, were seen at Toronto, in the afternoon of the next day—"not seen clearly enough to determine their nature, but they did not seem to be clouds or birds or smoke, and it was suggested that they were airships cruising over the city." *Toronto Daily Star*, February 10 —"They passed from west to east, in three groups, and then returned west in more scattered formation, about seven or eight in all."

36

August, 1914—this arena-like earth, with its horizon banking high into a Coliseum, when seen from not too far above—faint, rattling sounds of the opening of boundaries—tawny formations slinking into the arena—their crouchings and seizures and crunchings. Aug. 13, 1914—things that were gathering in the sky. They were seen by G. W. Atkins, of Elstree, Herts, and were seen again upon the 16th and the 17th (*Observatory*, 37-358). Sept. 9, 1914—a host in the sky; watched several hours by W. H. Steavenson (*Jour. B. A. A.*, 25-27). There were round appearances, but some of them were shaped like dumbbells. They were not seeds, snowflakes, insects, nor anything else that they "should" have been, according to Mr. Steavenson. He says that they were large bodies.

Oct. 10, 1914—a ship that was seen in the sky—or "an absolutely black, spindle-shaped object" crossing the sun. It was seen, at Manchester, by Albert Buss (*Eng. Mec.*, 100-236). "Its extraordinarily clear-cut outline was surrounded by a kind of halo, giving the impression of a ship, plowing her way through the sea, throwing up white-foamed waves with her prow."

Mikkelsen (*Lost in the Arctic*, p. 345):

"During the last few days (October, 1914) we have been much tumbled up and down in our minds, owing to a remarkable occurrence, somewhat in the nature of Robinson Crusoe's encounter with the footprints in the sand. Our advance load has been attacked—an empty petroleum cask is found, riddled with tiny holes, such as would be made by a charge of shot! Now a charge of shot is scarcely likely to materialize out of nowhere; one is accustomed to associate the phenomenon with the presence of human beings. It is none of our doing— then whose doing is it? We hit upon the wildest theories to account for it, as we sit in the tent, turning the mysterious object over and over. No beast of our acquaintance could make all those little round holes: what animal could even open its jaws so wide? And why should anybody take the trouble to make a target of our gear? Are there Eskimos about—Eskimos with guns? There are no footprints to be seen: it could scarcely have been an animal—the whole thing is highly mysterious."

Jan. 31, 1915—a symbolic-looking formation upon the moon—six or seven white spots, in Littrow, arranged like the Greek letter *Gamma* (*Eng. Mec.*, 101-47).

Feb. 13, 1915—Steep Island, Chusan Archipelago—a lighthouse-keeper complained to Capt. W. F. Tyler, R.N., that a British warship had fired a projectile at the lighthouse. But no vessel had fired a shot, and it is said that the object must have been a meteor (*Nature*, 97-17).

In the middle of February, 1915, the planet Venus was about two months and a half past inferior conjunction. If objects like navigating constructions were seen in the sky, at this time, there may be an association, but I am turning against that association, feeling that it is harmful to our wider expression that extra-mundane vessels have been seen in the sky of this earth, and that they come from regions at present unknown. *New York Tribune*, Feb. 15, 1915— that, at 10 P.M., February 14, three aëroplanes had been seen to cross the St. Lawrence river, near Morristown, N. Y., according to reports, but that, in the opinion of the Dominion police, nothing but fire-balloons had been seen. It is said that two "responsible residents" had seen two of the objects cross the river, between 8 and 8:30 P.M., and then return five hours later. In the Canadian Parliament, Sir Wilfred Laurier had said that, at 9 P.M., he had been called up by the Mayor of Brockwell, telling him that three aëroplanes with "powerful searchlights" had crossed the St. Lawrence. The story is told in the *New York Herald*. Here it is said that, according to the Chief of Police, of Ogdensburg, N.Y., a farmer, living five miles from Ogdensburg, had reported having seen an aëroplane, upon the 12th. Then it is said that the mystery had been solved: that, while celebrating the one hundredth anniversary of peace between the United States and Canada, some young men of Morristown had sent up paper balloons, which had exploded in the sky, after 9 P.M., night of the 14th. *New York Times*—that the objects had been seen first at Guananoque, Ontario. Here it is said that the balloon-story is absurd. According to the Dominion Observatory, the wind was, at the time, blowing from the east, and the objects had traveled toward the northeast. It is said that one of the objects had, for several minutes, turned a powerful searchlight upon the town of Brockwell.

Upon Dec. 11, 1915, Bernard Thomas, of Glenorchy, Tasmania, saw a "particularly bright spot upon the moon" (*Eng. Mec.*, 103-10). It was on the north shore of the Mare Crisium, and "looked almost like a star." In Dr. Thomas' opinion, it was sunlight reflected from the rim of a small crater. The crater Picard is near the north shore of the Mare Crisium, and most of the illuminations near Picard have occurred several months from an opposition of Mars.

In December, 1915, another new formation upon the moon—reported from the Observatory of Paris—something like a black wall from the center to the ramparts of Aristillus (*Bull. Soc. Astro. de France*, 30-383).

Jan. 12, 1916—a shock in Cincinnati, Ohio. Buildings were shaken. The quake was from an explosion in the sky. Flashes were seen in the sky. (*New York Herald*, Jan. 13, 1916.)

Feb. 9, 1916—opposition of Mars.

In the *English Mechanic*, 104-71, James Ferguson writes that someone had seen, at 11 o'clock, night of July 31, 1916, at Ballinasloe, Ireland, just such a moving thing, or just such a sailing, exploring thing as is now familiar in our records. For fifteen minutes it moved in a northwesterly direction. For three quarters of an hour it was stationary. Then it moved back to the point where first it had been seen, remaining visible until four o'clock in the morning. Whatever this object may have been, it left the sky at about the time that Venus appeared, as a "morning star," in the sky at Ballinasloe, and resembles the occurrence of Sept. 11, 1852, reported by Lord Wrottesley. Inferior conjunction of Venus was upon July 3, 1916. We have noticed that all occurrences that we somewhat reluctantly associate with nearness of Venus associate more with times of greatest brilliance, five weeks before and after inferior conjunction, than with dates of conjunction. Somebody may demonstrate that at these times Venus comes closest to this earth.

Oct. 10, 1916—a reddish shadow that spread over part of the lunar crater Plato; reported from the Observatory of Florence, Italy (*Sci. Amer.*, 121-181).

Nov. 25, 1916—about twenty-five bright flashes, in rapid succession, in the sky of Cardiff, Wales, according to Arthur Mee (*Eng. Mec.*, 104-239).

Col. Markwick writes, in the *Jour. B. A. A.*, 27-188, that, at 6:10 P.M., April 15, 1917, he had seen, upon the sun, a solitary spot, different from all sunspots that he had seen in an experience of forty-three years. Col. Markwick had written to Mr. Maunder, of the Greenwich Observatory, and had been told that, in photographs taken of the sun upon this day, one at 11:17 and another at 11:20 o'clock, there was no sign of a sunspot.

July 4, 1917—an eclipse of the sun, and an extraordinary luminous object said to have been a meteor, in France (*Bull. Soc. Astro. de France*, 31-299). About 6:20 P.M., this day, there was an explosion over the town of Colby, Wisconsin, and a stone fell from the sky (Science, Sept. 14, 1917).

Aug. 29, 1917—a luminous object that was seen moving upon the moon (*Bull. Soc. Astro. de France*, 31-439).

Feb. 21, 1919—an intensely black line extending out from the lunar crater Lexall (*Eng. Mec.*, 109-517).

Upon May 19, 1919, while Harry Hawker was at sea, untraceable messages, meaningless in the languages of this earth, were picked up by wireless, according to dispatches to the newspapers. They were interpreted as the letters *K U J* and *V K A J.*

In October, 1913, occurred something that may not be so very mysterious because of nearness to the sea. One supposes that if extra-mundane vessels have sometimes come close to this earth, then sailing away, terrestrial aeronauts may have occasionally left this earth, or may have been seized and carried away from this earth. Upon the morning of Oct. 13, 1913, Albert Jewel started to fly in his aeroplane from Hempstead Plains, Long Island, to Staten Island. The route that he expected to take was over Jamaica Bay, Brooklyn, Coney Island, and the Narrows. *New York Times*, Oct. 14, 1913—"That was the last seen or heard of him ... he has been as completely lost as if he had evaporated into air." But as to the disappearance of Capt. James there are circumstances that do call for especial attention. *New York Times*, June 2, 1919 —that Capt. Mansell R. James was lost somewhere in the Berkshire Hills, upon his flight from Boston to Atlantic City, or, rather, upon the part of his route between Lee, Mass., and Mitchel Field, Long Island. He had left Lee upon May 29th. Over the Berkshires, or in the Berkshires, he had disappeared. According to later dispatches, searching parties had "scoured" the Berkshires, without finding a trace of him. Upon June 4th, army planes arrived and searched systematically. There was general excitement, in this mystery of Capt. James. Rewards were offered; all subscribers of the Southern New England Telephone Company were enlisted in a quest for news of any kind; boy scouts turned out. Up to this date of writing there has been nothing but a confusion of newspaper dispatches: that two children had seen a plane, about thirteen miles north of Long Island Sound; that two men had seen a plane fall into the Hudson River, near Poughkeepsie; that, in a gully of Mount Riga, near Millerton, N. Y., had been found the remains of a plane; that part of a plane had been washed ashore from Long Island Sound, near Branford. The latest interest in the subject that I know of was in the summer of 1921. A heavy object was known to be at the bottom of the Hudson River, near Poughkeepsie, and was thought to be Capt. James' plane. It was dredged up and found to be a log.

For an extraordinary story of windows, in Newark, N. J., that were perforated by unfindable bullets, see *New York Evening Telegram*, Sept. 19, 1919, and the *Newark Evening News*. The occurrence is a counterpart of Mikkelsen's experience.

The detonations at Reading were heard seven years apart. Here it is not quite seven years later. London *Times*, Sept. 26, 1919—that upon September 25, a shock had been felt at Reading; that inquiries had led to information of no known explosion near Reading. In the Times, October 14, Mr. H. L. Hawkins writes that the shock was "quite definitely an earthquake, but its origin was superficial" and that the shock "was transmitted through the earth more than through the air." In the London *Daily Chronicle*, September 27, Mr. Hawkins, having considered all suggestions that the shock was a subterranean earthquake, had written: "However, as the whole thing terminated in a bump and a big bang, without subsequent shaking of the ground, it points more to an explosion of a natural type up in the air than to a real earthquake." And, in the London *Daily Mail*, Mr. Hawkins is quoted: that if the detonation were local, he would believe that it was an aërial explosion ("meteoric"); but, if it were widespread, it would be considered an earthquake. And in the whole series of the Reading phenomena, this violent detonation was most distinctly local to Reading.

Reading Observer, Sept. 27, 1919—"The most probable explanation of the occurrence is that there was an explosion somewhere near enough to affect the town.... Officials at the Greenwich Observatory were unable to throw any light on the matter, and said that their instruments showed no signs of earth-disturbance."

It is said that the sound and shock were violent, and that, in the residential parts of Reading, the streets were crowded with persons discussing the occurrence.

There was a similar shock in Michigan, Nov. 27, 1919. In many cities, persons rushed from their homes, thinking that there had been an earthquake (*New York Times*, November 28). But, in Indiana, Illinois, and Michigan, a "blinding glare" was seen in the sky. Our acceptance is that this occurrence is, upon a small scale, of the type of many catastrophes in Italy and South America, for instance, when just such "blinding glares" have been seen in the sky, data of which have been suppressed by conventional scientists, or data of which have not Impressed conventional scientists.

English Mechanic, 110-257—J. W. Scholes, of Huddersfield, writes that, upon Dec. 19, 1919, he saw, near the lunar crater Littrow, "a, very conspicuous black-ink mark." Upon page 282, W. J. West, of Gosport, writes that he had seen the mark upon the 7th of December.

March 22, 1920—a light in the sky of this earth, and an illumination upon the moon (*Eng. Mec.*, III-142). That so close to this earth is the moon that illuminations known as "auroral" often affect both this earth and the moon.

July 20 and 21, and Sept. 13, 1920—dull rumbling sounds and quakes at Comrie, Perthshire (London *Times*, July 23 and Sept. 14, 1920).

According to a dispatch to the *Los Angeles Times*—clipping sent to me by Mr. L. A. Hopkins, of Chicago—thunder and lightning and heavy rain, at Portland, Oregon, July 21, 1920: objects falling from the sky; glistening, white fragments that looked like "bits of polished china." "The explanation of the local Weather Bureau is that they may have been picked up by a whirlwind and carried to the district where they were found." The objection to this standardized explanation is the homogeneousness of the falling objects. How can one conceive of winds raging over some region covered with the usual great diversity of loose objects and substances, having a liking for little white stones, sorting over maybe a million black ones, green ones, white ones, and red ones, to make the desired selection? One supposes that a storm brought to this earth fragments of a manufactured object, made of something like china, from some other world.

In the *Literary Digest*, Sept. 2, 1921, is published a letter from Carl G. Gowman, of Detroit, Michigan, upon the fall from the sky, in southwest China, Nov. 17 (1920?) of a substance that resembled blood. It fell upon three villages close together, and was said to have fallen somewhere else forty miles away. The quantity was great: in one of the villages, the substance "covered the ground completely." Mr. Gowman accepts that this substance did fall from the sky, because it was found upon roofs as well as upon the ground. He rejects the conventional red-dust explanation, because the spots did not dissolve in several subsequent rains. He says that anything like pollen is out of the question, because at the time nothing was in bloom.

Nov. 23, 1920—a correspondent writes, to the *English Mechanic*; 112-214, that he saw a shaft of light projecting from the moon, or a spot so bright that it appeared to project, from the limb of the moon, in the region of Funerius.

About Jan. 1, 1921—several irregular, black objects that crossed the sun. To the Rev. William Ellison (*Eng. Mec.*, 112-276) they looked like pieces of burnt paper.

July 25, 1921—a loud report, followed by a sharp tremor, and a rumbling sound, at Comrie (London *Times*, July 27, 1921).

July 31, 1921—a common indication of other lands from which come objects and substances to this earth—but our reluctance to bother with anything so ordinarily marvelous

Because we have conceived of intenser times and furies of differences of potential between this earth and other worlds: torrents of dinosaurs, in broad volumes that were streaked with lesser animals, pouring from the sky, with a foam of tusks and fangs, enveloped in a bloody vapor that was falsely dramatized by the sun, with rainbow-mockery. Or, in terms of planetary emotions, such an outpouring was the serenade of some other world to this earth. If poetry is imagery, and if a flow of images be solid poetry, such a recitation was in three-dimensional hyperbole that was probably seen, or overheard, and criticized in Mars, and condemned for its extravagance in Jupiter. Some other world, meeting this earth, ransacking his solid imagination and uttering her living metaphors: singing a flood of mastodons, purring her butterflies, bellowing an ardor of buffaloes. Sailing away—sneaking up close to the planet Venus, murmuring her antelopes, or arching his periphery and spitting horses at her—

Poor, degenerate times—nowadays something comes close to this earth and lisps little commonplaces to her—

July 31, 1921—a shower of little frogs that fell upon Anton Wagner's farm, near Stirling, Conn. (*New York Evening World*, Aug. I, 1921).

At sunset, Aug. 7, 1921, an unknown luminous object was seen, near the sun, at Mt. Hamilton, by an astronomer, Prof. Campbell, and by one of those who may some day go out and set foot upon regions that are supposed not to be: by an aviator, Capt. Rickenbacker. In the *English Mechanic*, 114-211, another character in these fluttering vistas of the opening of the coming drama of Extra-geography, Col. Markwick, a conventional astronomer and also a recorder of strange things, lists other observations upon this object, the earliest upon the 6th, by Dr. Emmert, of Detroit. In the *English Mechanic*, 114-241, H. P. Hollis, once upon a time deliciously "exact" and positive, says something, in commenting upon these observations, that looks like a little weakness in Exclusionism, because the old sureness is turning slightly shaky—"that there are more wonderful things in the sky than we suspect, or that it is easy to be self-deceived."

It is funny to read of an "earthquake," described in technical lingo, and to have a datum that indicates that it was no earthquake at all, in the usual seismologic sense, but a concussion from an explosion in the sky. Aug. 7, 1921—a severe shock at New Canton, Virginia. See *Bull. Seis. Soc. Amer.*, 11-197—Prof. Stephen Taber's explanation that the shock had probably originated in the slate belt of Buckingham County, intensity about V on the R.-F. scale. But then it is said that, according to the "authorities" of the McCormick Observatory, the concussion was from an explosion in the sky. The time is coming when nothing funny will be seen in this subject, if some day be accepted at least parts of the masses of data that I am now holding back, until I can more fully develop them—that some of the greatest catastrophes that have devastated the face of this earth have been concussions from explosions in the sky, so repeating in a local sky weeks at a time, months sometimes, or intermittently for centuries, that fixed origins above the ravaged areas are indicated.

New York Tribune, Sept. 2, 1921:

"J. C. H. Macbeth, London Manager of the Marconi Wireless Telegraph Company, Ltd., told several hundred men, at a luncheon of the Rotary Club, of New York, yesterday, that Signor Marconi believed he had intercepted messages from Mars, during recent atmospheric experiments with wireless on board his yacht Electra, in the Mediterranean. Mr. Macbeth said that Signor Marconi had been unable to conceive of any other explanation of the fact that, during his experiments, he had picked up magnetic wavelengths of 150,000 meters, whereas the maximum length of wave-production in the world today is 14,000 meters. The regularity of the signals, Mr. Macbeth declared, disposed of any assumption that the waves might have been caused by electrical disturbance. The signals were unintelligible, consisting apparently of a code, the speaker said, and the only signal recognized was one resembling the letter V in the Marconi code." See datum of May 19, 1919. But, in the summer of 1921, the planet Mars was far from opposition. The magnetic vibrations may have come from some other world. They may have had the origin of the sounds that have been heard at regular intervals—

The San Salvadors of the sky—

And we return to the principle that has been our re-enforcement throughout: that existence is infinite serialization, and that, except in particulars, it repeats
—

That the dot that spread upon the western horizon of Lisbon, March 4, 1493, cannot be the only ship that comes back from the unknown, cargoed with news
—

And it may be September this, nineteen hundred and twenty or thirty something, or February that, nineteen hundred and twenty or thirty something else—and, later, see record of it in *Eng. Mec.*, or *Sci. Amer.*, vol. and p. something or another—a speck in the sky of this earth—the return of somebody from a San Salvador of the sky—and the denial by the heavens themselves, which may answer with explosions the vociferations below them, of false calculations upon their remotenesses. If the heavens do not participate with snow, the skyscrapers will precipitate torn up papers and shirts and skirts, too, when the papers give out.

There will be a procession. Somebody will throw little black pebbles to the crowds. Over his procession will fly blue-fringed cupids. Later he will be insulted and abused and finally hounded to his death. But, in that procession, he will lead by the nose an outrageous thing that should not be: about ten feet long, short-winged, waddling on webbed feet. Insult and abuse and death—he will snap his fingers under the nose of the outrageous thing. It will be worth a great deal to lead that by the nose and demonstrate that such things had been seen in the sky, though they had been supposed to be angels. It will be a great moment for somebody. He will come back to New York, and march up Broadway with his angel.

Some now unheard-of De Soto, of this earth, will see for himself the Father of Cloudbursts.

A Balboa of greatness now known only to himself will stand on a ridge in the sky between two auroral seas.

Fountains of Everlasting Challenge.

Argosies in parallel lines and rabbles of individual adventurers. Well enough may it be said that they are seeds in the sky. Of such are the germs of colonies.

THAT the Geo-system is an incubating organism, of which this earth is the nucleus—but an organism that is so strongly characterized by conditions and features of its own that likening it to any object internal to it is the interpreting of a thing in terms of a constituent—so that we think of an organism that is incompletely, or absurdly inadequately, expressible in terms of the egg-like and the larval and other forms of the immature—a geo-nucleated system that is dependent upon its externality as, in one way or another, is every similar, but lesser and included, thing—stimulated by flows of force that are now said to be meteoric, though many so-called "meteoric" streams seem more likely to be electric, that radiate from the umbilical channels of its constellations—vitalized by its sun, which is itself replenished by the comets, which, coming from external reservoirs of force, impart to the sun their freightages, and, unaffected by gravitation, return to an external existence, some of them even touching the sun, but showing no indication of supposed solar attraction.

In a technical sense we give up the doctrine of Evolution. Ours is an expression upon Super-embryonic Development, in one enclosed system. Ours is an expression upon Design underlying and manifesting in all things within this one system, with a Final Designer left out, because we know of no designing force that is not itself the product of remoter design. In terms of our own experience we cannot think of an ultimate designer, any more than we can think of ultimacy in any other respect. But we are discussing a system that, in our conception, is not a final entity; so then no metaphysical expression upon it is required.

I point out that this expression of ours is not meant for aid and comfort to the reactionaries of the type of Col. W. J. Bryan, for instance: it is not altogether anti-Darwinian: the concept of Development replaces the concept of Evolution, but we accept the process of Selection, not to anything loosely known as Environment, but relatively to underlying Schedule and Design, predetermined and supervised, as it were, but by nothing that we conceive of in anthropomorphic terms.

I define what I mean by dynamic design, in the development of any embryonic thing: a pre-determined, or not accidental, or not irresponsible, passage along a schedule of phases to a climax of unification of many parts. Some of the aspects of this process are the simultaneous varying of parts, with destiny, and not with independence, for their rule, or with future co-ordinations and functions for their goal; and their survival while still incipient, not because they are fittest relatively to contemporaneous environment, so not because of usefulness or advantage in the present, inasmuch as at first they are not only functionless but also discordant with established relations, but surviving because they are in harmony with the dynamic plan of a whole being: and the presence of forces of suppression, or repression, as well as forces of stimulation and protection, so that parts are held back, or are not permitted to develop before their time.

If we accept that these circumstances of embryonic development are the circumstances of all wider development, within one enclosed system, the doctrine of Darwinian Evolution, as applied generally, will, in our minds, have to be replaced by an expression upon Super-embryonic Development, and Darwinism, unmodified, will become to us one more of the insufficiencies of the past. Darwinism concerns itself with the adaptations of the present, and does heed the part that the past has played, but, in Darwinism, there is no place for the influence of the future upon the present.

Consider any part of an embryonic thing—the heart of an embryo—and at first it is only a loop. It will survive, and it will be nourished in its functionless incipiency; also it will not be permitted to become a fully developed heart before its scheduled time arrives; its circumstances are dominated by what it will be in the future. The eye of an embryo is a better instance.

Consider anything of a sociologic nature that ever has grown: that there never has been an art, science, religion, invention that was not at first out of accord with established environment, visionary, preposterous in the light of later standards, useless in its incipiency, and resisted by established forces so that, seemingly animating it and protectively underlying it, there may have been something that in spite of its unfitness made it survive for future usefulness. Also there are data for the acceptance that all things, in wider being, are held back as well as protected and prepared for, and not permitted to develop before comes scheduled time. Langley's flying machine makes me think of something of the kind—that this machine was premature; that it appeared a little before the era of aviation upon this earth, and that therefore Langley could not fly. But this machine was capable of flying, because, some years later, Curtis did fly in it. Then one thinks that the Wright Brothers were successful, because they did synchronize with a scheduled time. I have heard that it is questionable that Curtis made no alterations in Langley's machine. There is no lack of instances. One of the greatest of secrets that have eventually been found out was for ages blabbed by all the pots and kettles in the world—but that the secret of the steam engine could not, to the lowliest of intellects, or to supposititiously highest of intellects, more than adumbratively reveal itself until came the time for its co-ordination with the other phenomena and the requirements of the Industrial Age. And coal that was stored in abundance near the surface of the ground—and the needs of dwellers over coal mines, veins of which were often exposed upon the surface of the ground, for fuel—but that this secret, too, obvious, too, could not be revealed until the coming of the Industrial Age. Then the building of factories, the inventing of machines, the digging of coal, and the use of steam, all appearing by simultaneous variation, and co-ordinating, Shores of North America—nowadays, with less hero-worship than formerly, historians tell us that, to English and French fishermen, the coast of Newfoundland was well-known, long before the year 1492; nevertheless, to the world in general, it was not, or, according to our acceptances, could not be, known. About the year 1500, a Portuguese fleet was driven by storms to the coast of Brazil, and returned to Europe. Then one thinks that likely enough, before the year 1492, other vessels had been so swept to the coasts of the western hemisphere, and had returned—but that data of westward lands could not emerge from the suppressions of that era—but that the data did survive, or were preserved for future usefulness—that there are "Thou shalt nots" engraved upon something underlying all things, and then effacing, when phases pass away.

We conceive now of all buildings—within one enclosed system—in terms of embryonic building, and of all histories as local aspects of Super-embryonic Development. Cells of an embryo build falsely and futilely, in the sense that what they construct will be only temporary and will be out of adjustment later. If, however, there are conditions by which successive stages must be traversed before the arrival of maturity, ours is an expression upon the functioning of the false and the futile, in which case these terms, as derogations, should not be applied. We see that the cells that build have no basis of their own; that for their formations there is nothing of reason and necessity of their own, because they flourish in other formations quite as well. We see that they need nothing of basis, nor of guidance of their own, because basis and guidance are of the essence of the whole. All are responses, or correlates, to a succession of commandments, as it were, or of dominant, directing, supervising spirits of different eras: that they take on appearances that are concordant with the general gastrula era, changing when comes the stimulus to agree with the reptilian era, and again responding harmoniously when comes the time of the mammalian era. It is in accordance with our experience that never has human mind, scientific, religious, philosophic, formulated one basic thought, one finally true law, principle, or major premise from which guidance could be deduced. If any thought were true and final it would include the deduced. We conceive that there has been guidance, just the same, if human beings be conceived of as cellular units in one developing organism; and that human minds no more need foundations of their own than need the sub-embryonic cells that build so preposterously, according to standards of later growth, but build as they are guided to build. In this view, human reason is tropism, or response to stimuli, and reasoning is the trial-and-error process of the most primitive unicellular organisms, a susceptibility to underlying mandates, then a groping in perhaps all possible distortions until adjustment with underlying requirements is reached. In this view, then, though there are, for instance, no atoms in the Daltonian sense, if in the service of a building science, the false doctrine of the atoms be needed, the mind that responds, perhaps not to stimulus, but to requirement, which seems to be a negative stimulus, and so conceives, is in adjustment and reaches the state known as success. I accept, myself, that there may be Final Truth, and that it may be attainable, but never in a service that is local or special in any one science or nation or world.

It is our expression that temporary isolations characterize embryonic growth and super-embryonic growth quite as distinctly as do expansions and co-ordinations. Local centers of development in an egg—and they are isolated before they sketch out attempting relations. Or in wider being—hemisphere isolated from hemisphere, and nation from nation—then the breaking down of barriers—the appearance of Japan out of obscurity—threads of a military plasm are cast across an ocean by the United States.

Shafts of light that have pierced the obscurity surrounding planets—and something like a star shines in Aristarchus of the moon. Embryonic heavens that have dreamed—and that their mirages will be realized some day. Sounds and an interval; sounds and the same interval; sounds again—that there is one integrating organism and that we have heard its pulse.

38

Feb. 7, 1922—an explosion "of startling intensity" in the sky of the northwestern point of the London Triangle (*Nature*, Feb. 23, 1922).

Repeating phenomena in a local sky—in *L'Astronomie*, 36-201, it is said that, at Orsay (Seine-et-Oise), Feb. 15, 1922, a detonation was heard in the sky, and that 9 hours later a similar sound was heard, and that an illumination was seen in the sky. It is said that, 10 nights later, at Verneuil, in the adjoining province, Oise, a great, fiery mass was seen falling from the sky.

March 12, 1922—rocks that had been falling "from the clouds," for three weeks, at Chico, a town in an "earthquake region" in California (*New York Times*, March 12, 1922). Large, smooth rocks that "seemed to come straight from the clouds."

In the *San Francisco Chronicle*, in issues dating from the 12th to the 18th of March—clippings sent to me by Mr. Maynard Shipley, writer and lecturer upon scientific subjects, if there be such subjects—the accounts are of stones that, for four months, had been falling intermittently from the sky, almost always upon the roofs of two adjoining warehouses, in Chico, but, upon one occasion, falling three blocks away: "a downpour of oval-shaped stones"; "a heavy shower of warm rocks." *San Francisco Call*, March 16—"warm rocks."

It is said that crowds gathered, and that upon the 17th of March a "deluge" of rocks fell upon a crowd, injuring one person. The police "combed" all surroundings: the only explanation that they could think of was that somebody was firing stones from a catapult. One person was suspected by them, but, upon the 14th of March, a rock fell when he was known not to he in the neighborhood.

The circumstances point to one origin of these stones, stationary in the sky, above the town of Chico.

Upon the first of January, 1922, the attention of Marshal J. A, Peck, of Chico, had been called to the phenomena. After investigating more than two months, he said (*San Francisco Examiner*, March 14) "I could find no one through my investigations who could explain the matter. At various times I have heard and seen the stones. I think someone with a machine is to blame."

Prof. C. K. Studley, vice-president of the Teachers' College, Chico, is quoted in the *Examiner:*

"Some of the rocks are so large that they could not be thrown by any ordinary means. One of the rocks weighs 16 ounces. They are not of meteoric origin, as seems to have been hinted, because two of them show signs of cementation, either natural or artificial, and no meteoric factor was ever connected with a cement factory."

Once upon a time, dogmatists supposed, asserted, angrily declared sometimes, that all stones that fall from the sky must be of "true meteoric material." That time is now of the past. See *Nature*, 105-759—a description of two dissimilar stones, cemented together, seen to fall from the sky, at Cumberland Falls, Ky., April 9, 1919.

Miriam Allen de Ford (P. O. Box 573, San Francisco, Cal.—or see the *Readers' Guide*) has sent me an account of her own observations. About the middle of March, 1922, she was in Chico, and investigated. Went to the scene of the falling rocks; discussed the subject with persons in the crowd. "While I was discussing it with some bystanders, I looked up at the cloudless sky, and suddenly saw a rock falling straight down, as if becoming visible when it came near enough. This rock struck the roof with a thud, and bounced off on the track beside the warehouse, and I could not find it." "I learned that the rocks had been falling since July, 1921, though no publicity arose until November."

There have been other phenomena at Chico. In the *New York Times*, Sept. 2, 1878, it is said that, upon the 20th of August, 1878, according to the *Chico Record*, a great number of small fishes fell from the sky, at Chico, covering the roof of a store, and falling in the streets, upon an area of several acres. Perhaps the most important observation is that they fell from a cloudless sky. Several occurrences are listed as earthquakes, by Dr. Holden, in his Catalogue; but the detonations that were heard at Oroville, a town near Chico, Jan. 2, 1887, are said, in the *Monthly Weather Review*, 1887-24, to have been in the sky. Upon the night of March 5-6, 1885, according to the *Chico Chronicle*, a large object, of very hard material, weighing several tons, fell from the sky, near Chico (*Monthly Weather Review*, March, 1885). In the year 1893, an iron object, said to be meteoritic, was found at Oroville (*Mems. Nat. Acad. Sci.*, 13-345).

My own idea is either that there is land over the town of Chico, and not far away, inasmuch as objects from it fall with a very narrow distribution, or that far away, and therefore invisible, there may be land from which objects have been carried in a special current to one very small part of this earth's surface. If anyone would like to read an account of stones that fell intermittently for several days, clearly enough as if in a current, or in a field of special force, of some kind, at Livet, near Clavaux, France, December, 1842, see the London *Times*, Jan. 13, 1843. There have been other such occurrences. Absurdly, when they were noticed at all, they were supposed to be psychic phenomena. I conceive that there is no more of the psychic to these occurrences than there is to the arrival of seeds from the West Indies upon the coast of England. Stones that fell upon a house, near the Pantheon, Paris, for three weeks, January, 1849—see Dr. Wallace's *Miracles and Modern Spiritualism*, p. 284. Several times, in the course of this book, I have tried to be reasonable. I have asked what such repeating phenomena in one local sky do indicate, if they do not indicate fixed origins in the sky. And if such occurrences, supported by many data in other fields, do not indicate the stationariness of this earth, with new lands not far away—tell me what it is all about. The falling stones of Chico—new lands in the sky—or what?

Boston Transcript, March 21, 1922—clipping sent to me by Mr. J. David Stern, Editor and Publisher of the *Camden* (N. J.)*Daily Courier*—

"Geneva, March 21—During a heavy snow storm in the Alps recently thousands of exotic insects resembling spiders, caterpillars, and huge ants fell on the slopes and quickly died. Local naturalists are unable to explain the phenomenon, but one theory is that the insects were blown in on the wind from a warmer climate."

The fall of unknown insects in a snow storm is not the circumstance that I call most attention to. It is worth noting that I have records of half a dozen similar occurrences in the Alps, usually about the last of January, but the striking circumstance is that insects of different species and of different specific gravities fell together. The conventional explanation is that a wind, far away, raised a great variety of small objects, and segregated them according to specific gravity, so that twigs and grasses fell in one place, dust some other place, pebbles somewhere else, and insects farther along somewhere. This would be very fine segregation. There was no very fine segregation in this occurrence. Something of a seasonal, or migratory, nature, from some other world, localized in the sky, relatively to the Alps, is suggested.

May 4, 1922—discovery, by F. Burnerd, of three long mounds in the lunar crater Archimedes. See the *English Mechanic*, 115-194, 218, 268, 278. It seems likely that these constructions had been recently built.

St. Thomas, Virgin Islands, May 18, 1922 (Associated Press)—particles of matter falling continuously for several days. "The phenomenon is supposed here to be of volcanic origin, but all the volcanoes of the West Indies are reported as quiet."

New York Tribune, July 3, 1922—that, for the fourth time in one month, a great volume of water, or a "cloudburst," had poured from one local sky, near Carbondale, Pa.

Oct. 15, 1922—a large quantity of white substance that fell upon the shores of Lake Michigan, near Chicago. It fell upon the clothes of hundreds of persons, fell upon the campus of Northwestern University, likely enough fell upon the astronomical observatory of the University. It occurred to one of these hundreds, or thousands, of persons to collect some of this substance. He is Mr. L. A. Hopkins, 111 West Jackson Blvd., Chicago. He sent me a sample. I think that it is spider web, because it is viscous: when burned it chars with the crinkled effect of burned hair and feathers, and the odor is similar. But it is strong, tough substance, of a cottony texture, when rolled up. The interesting circumstance to me is that similar substance has fallen frequently upon this earth, in October, but that, in terrestrial terms, seasonal migrations of aeronautical spiders cannot be thought of, because in the tropics and in Australia, as well as in the United States and in England, such showers have occurred in October. Then something seasonal, but seasonal in an extra-mundane sense, is suggested. See the *Scientific Australian*, September, 1916—that, from October 5 to 29, 1915, an enormous fall of similar substance occurred upon a region of thousands of square miles, in Australia.

Time after time, in data that I have only partly investigated, occur declarations that, during devastations commonly known as "earthquakes," 'in Chile, the sky has flamed, or that "strange illuminations" in the sky have been seen. In the *Bull. Seis. Soc. Amer.*, for instance, some of these descriptions have been noted, and have been hushed up with the explanation that they were the reports of unscientific persons.

Latest of the great quakes in Chile—1,500 dead "recovered" in one of the cities of the Province of Atacama. *New York Tribune*, Nov. 15, 1922—"Again, today, severe earthquakes shook the Province of Coquimbo and other places, and strange illuminations were observed over the sea, off La Serena and Copiapo."

Back to Crater Mountain, Arizona, for an impression—but far more impressive are similar data as to these places of Atacama and Copiapo, in Chile. In the year 1845, M. Darlu, of Valparaiso, read, before the French Academy, a paper, in which he asserted that, in the desert of Atacama, which begins at Copiapo, meteorites are strewn upon the ground in such numbers that they are met at every step. If these objects fell all at one time in this earthquake region, we have another instance conceivably of mere coincidence between the aërial and the seismic. If they fell at different times, the indications are of a fixed relationship between this part of Chile and a center somewhere in the sky of falling objects commonly called "meteorites" and of cataclysms that devastate this part of Chile with concussions commonly called "earthquakes." There is a paper upon this subject in Science, 14-434. Here the extreme abundance asserted by M. Darlu is questioned: it is said that only thirteen of these objects were known to science.

But, according to descriptions, four of them are stones, or stone-irons, differing so that, in the opinion of the writer, and not merely so interpreted by me, these four objects fell at different times. Then the nine others are considered. They are nickel-irons. They, too, are different, one from another. So then it is said that these thirteen objects, all from one place, were, with reasonable certainty, the products of different falls.

Behind concepts that sometimes seem delirious, I offer—a reasonable certainty—

That, existing somewhere beyond this earth, perhaps beyond a revolving shell in which the nearby stars are openings, there are stationary regions, from which, upon many 'occasions, have emanated "meteors," sometimes exploding catastrophically over Atacama, Chile, for instance. Coasts of South America have reeled, and the heavens have been afire. Reverberations in the sky—the ocean has responded with islands. Between sky and earth of Chile there have been flaming intimacies of destruction and slaughter and woe—

Silence that is conspiracy to hide past ignorance; that is imbecility, or that is the unawareness of profoundest hypnosis.

Hypnosis—

That the seismologists, too, have functioned in preserving the illusion of this earth's isolation, and by super-embryonic processes have been hypnotized into oblivion of a secret that has been proclaimed with avalanches of fire from the heavens, and that has babbled from books of the blood of crushed populations, and that is monumentalized in ruins.

———————————